Metamaterials for Antenna Applications

T0313189

Metamaterials for Antenna Applications

Amit K. Singh
Mahesh P. Abegaonkar
Shiban Kishen Koul

CRC Press
Taylor & Francis Group
Boca Raton London New York

CRC Press is an imprint of the
Taylor & Francis Group, an **Informa** business

First edition published 2022
by CRC Press
6000 Broken Sound Parkway NW, Suite 300, Boca Raton, FL 33487–2742

and by CRC Press
2 Park Square, Milton Park, Abingdon, Oxon, OX14 4RN

© 2022 Taylor & Francis Group, LLC

CRC Press is an imprint of Taylor & Francis Group, LLC

ISBN: 978-0-367-49350-9 (hbk)
ISBN: 978-0-367-49354-7 (pbk)
ISBN: 978-1-003-04588-5 (ebk)

DOI: 10.1201/9781003045885

Typeset in Palatino
by Apex CoVantage, LLC

Dedication

Sikhandar Prasad Singh, Sanjay Kumar Singh, Pushpa Devi,
Samta Kumari Singh, Ankit Dubey
Amit K. Singh

Nayana Abegaonkar, Rashmi Abegaonkar,
Arnav Abegaonkar, Pandurang Joshi, Madhavi Joshi
Mahesh P. Abegaonkar

Veena Koul, Mansi Narain, Grace Yeo Koul, Amendra Koul, Anant Koul
Shiban Kishen Koul

Contents

Preface

This book is the outcome of research work carried out at Microwave Lab, Centre for Applied Research in Electronics, Indian Institute of Technology Delhi, New Delhi, India, from 2014 to 2018. The authors adopted an engineering approach with systematic development of metamaterials with an emphasis on application of metamaterials and metasurfaces (MSs) for antenna applications.

The book presents the design and development of electromagnetic metamaterial and their application in microwave antennas. Chapter 1 introduces metamaterials with a brief historical background. The material properties and design of metamaterial utilizing a unit cell (UC) with wave interaction is discussed in this chapter. Chapter 2 discusses the fundamental idea of design and development of metamaterial structures and various measurement techniques. The studies carried out include design of wideband low-profile ultrathin MS for X-band application, miniaturization and bandwidth enhancement of MS, design and development of multiband ultrathin transmission and reflection-type MS for C-, X- and Ka-band applications. Chapter 3 presents the application of metamaterial for miniaturization of a microstrip patch antenna (MPA). The research work presented includes design of a dual-band highly miniaturized MPA with significant gain, design of a triple-band significant gain miniaturized patch antenna and a highly miniaturized triple-band MPA with significant measured gain in the broadside and backside. Chapter 4 extends the application of metamaterials to the design and development of high-gain antennas using reflection-type MS. Starting from the working principle of an open-air cavity resonator antenna, design of high-gain, high-aperture efficiency cavity resonator antenna for X-band application and design of cavity resonator antenna for C-band application are presented in this chapter. Chapter 5 discusses design and development of a high-gain antenna using a transmission-type MS. Detailed studies are reported on the working principle of the high-gain antenna using a transmission-type MS, design of high-gain MS lens antenna with wideband gain enhancement for C-band application and design of MS lens antenna for X-band application. Chapter 6 presents the design of the graded index metasurface (GIMS), including linear and angular/radial GIMS design, working principle of the GIMS lens antenna, design and development of linear graded index metasurface (LGIMS) lens antenna and radial graded index metasurface (RGIMS) lens antenna for beam steering and gain enhancement and design of widebeam steerable flattop beam antenna using GIMS lens. Chapter 7 discusses the design and development of microwave absorbers based on metamaterials, including the working principle of a microwave absorber, absorber measurement setup, design and characterization of a penta-band microwave absorber for S-, C-, X-, Ku- and Ka-band applications, design of triple-band polarization-insensitive ultrathin metamaterial absorber for S-, C- and X-band applications, design of a conformal double-band metamaterial absorber and design of a triple-band conformal polarization-insensitive metamaterial absorber.

The authors believe that the engineering approach of the design and development of metamaterials and their applications presented in this book will be extremely valuable for antenna designers who are interested in the development of novel antenna devices for future-generation communication systems, space and defence applications.

Authors

Amit K. Singh is Assistant Professor in the Department of Electrical Engineering, Indian Institute of Technology Jammu, India. He worked as a Postdoctoral Researcher in Korea Advanced Institute of Science and Technology, Daejeon, Republic of Korea, from November 2018 to February 2020. He completed his PhD at the Indian Institute of Technology Delhi, India, in 2018. Dr. Singh has authored and co-authored over 35 referred international journal and conference articles. He is the recipient of the International Union of Radio Science (URSI) Young Scientist Award (2021). He has been recognized by Wiley-USA as author of the Top Downloaded Paper in 2018–2019. He is the recipient of the Brain Korea–21 Fellowship 2019 for Excellence in Research. He is also recipient of the GCORE–2019 Fellowship by Global Center for Open Research with Enterprise, KAIST, South Korea. His current areas of research include metamaterials for 5G and future communications, millimeter wave high-gain beam steerable meta-antenna design, massive MIMO antenna for 5G and beyond, metasurface and antenna applications, deployable high-gain antenna for micro-satellite applications, metamaterial absorber and FSS, microwave imaging for biomedical applications and electromagnetic warfare.

Mahesh P. Abegaonkar completed his PhD in microwave sensors from the University of Pune, India, in 2002. He worked as Post-Doctoral Researcher and Assistant Professor (Contract) in the School of Electrical Engineering and Computer Science, Kyungpook National University, Daegu, South Korea, from June 2002–January 2005. In February 2005, he joined the Centre for Applied Research in Electronics (CARE), Indian Institute of Technology (IIT) Delhi, where he is currently an Associate Professor. He has been a Senior Member of IEEE and Young Associate of Indian National Academy of Engineering (INAE) since 2013. He was Secretary and Treasurer of the IEEE MTT-S Delhi Chapter during 2008–2017 and Vice-Chairman during 2017–2018. He was National Coordinator of Virtual Laboratories under Electronics and Communication Engineering Vertical, an initiative of the Ministry of Human Resource Development (MHRD), Govt. of India, under the National Mission on Education through Information and Communication Technologies (NMEICT).

He received the Young Engineer Award from INAE in 2008 in the field of RF and Microwave Engineering. He was also a recipient of a Senior Research

Fellowship from the Council for Scientific and Industrial Research (CSIR), Govt. of India, from 1998–2001. He regularly reviews manuscripts for IEEE and IET journals and letters.

He has supervised nine PHD students and is currently supervising three. He has also supervised 76 M.Tech. projects. He has over 160 publications in international journals and conferences and 35 publications in national conferences. His research has been cited by over 1,500 researchers from around the world. He has two international patents granted and one Indian patent filed. He has co-authored the book *Printed Resonant Periodic Structures and Their Applications*, published by CRC Press.

His research interests include periodic structures for antenna applications, millimetre wave antennas for 5G communications, MIMO antennas, printed antennas (reconfigurable, broadband, high gain, active, etc.).

Shiban Kishen Koul (Life Fellow, IEEE) received a BE degree in electrical engineering from the Regional Engineering College, Srinagar, India, in 1977, and M.Tech. and PhD degrees in microwave engineering from the Indian Institute of Technology Delhi, New Delhi, India, in 1979 and 1983, respectively.

He has been Emeritus Professor with the Indian Institute of Technology Delhi since 2019 and Mentor Deputy Director (Strategy and Planning, International Affairs) with IIT Jammu, Jammu and Kashmir, India, since 2018. He served as Deputy Director (Strategy and Planning) with IIT Delhi from 2012 to 2016. He also served as the Chairman of Astra Microwave Products Limited, Hyderabad, from 2009 to 2019 and Dr R. P. Shenoy Astra Microwave Chair Professor at IIT Delhi from 2014 to 2019. His research interests include: RF MEMS, high-frequency wireless communication, microwave engineering, microwave passive and active circuits, device modelling, millimeter wave IC design and reconfigurable microwave circuits and metamaterial-inspired MEMS and antennas for 5G applications.

He has successfully completed 38 major sponsored projects, 52 consultancy projects and 61 technology development projects. He has authored or co-authored 490 research papers, 13 state-of-the art books, four book chapters and two e-books. He holds 16 patents, six copyrights and one trademark. He has guided 25 PhD theses and more than 100 master's theses.

Prof. Koul is a Fellow of the Indian National Academy of Engineering, India, and Institution of Electronics and Telecommunication Engineers (IETE), India. He is the Chief Editor of *IETE Journal of Research* and Associate Editor of the *International Journal of Microwave and Wireless Technologies*, Cambridge University Press. He served as a Distinguished Microwave Lecturer of IEEE MTT-S for the period 2012–2014. He also served as an AdCom member of the IEEE MTT-S from 2010 to 2018 and is presently a member of the Awards, Nomination and Appointments, MGA, M&S and Education committees of the IEEE MTT-S. He is the recipient of numerous awards including IEEE MTT Society Distinguished Educator Award (2014); Teaching Excellence Award (2012) from IIT Delhi; Indian National Science Academy (INSA) Young Scientist Award (1986); Top Invention Award (1991) of the National Research Development Council for his contributions to the indigenous development of ferrite phase shifter technology; VASVIK Award (1994) for the development of Ka-band components and phase shifters; Ram Lal Wadhwa Gold Medal (1995) from the Institution of Electronics and Communication Engineers (IETE); Academic Excellence Award (1998) from the Indian Government for his pioneering contributions to phase control modules for Rajendra Radar, Shri Om Prakash Bhasin Award (2009) in the field of Electronics and Information Technology, VASVIK Award (2012) for the contributions made to the area of Information, Communication Technology (ICT) and M N Saha Memorial Award (2013) from IETE.

Abbreviations

ADS	advanced design system
AGIMS	angular graded index metasurface
AMC	artificial magnetic conductors
CLL	capacitive-loaded loops
CSRR	circular split-ring resonator
CsRR	circular slot ring resonator
CRR	circular ring resonator
CPW	coplanar waveguide
CRLH	composite right-/left-handed
CST	computer simulation technology
DASR	double annular slot ring resonator
DS	double stacked
DSDCRR	double-sided double circular ring resonator
DPS	double-positive
DNG	double-negative
EBG	electromagnetic bandgap
EM	electromagnetic
ENG	electrically negative
FSS	frequency-selective surfaces
FPC	Fabry–Perot cavity
GIMS	graded index metasurface lens
HPBW	half-power beam width
IDC	interdigital capacitor
LGIMS	linear graded index metasurface
LHM	left-handed material
MCSRR	modified circular split-ring resonator
MCsRR	modified circular slot ring resonator
MDCsRR	modified double circular slot ring resonator
MNG	magnetically negative
MS	metasurface
MTCsRR	modified triple circular slot ring resonator
MTCRR	modified triple circular ring resonator
NIM	negative-index metamaterial
NRW	Nicolson–Ross–Weir
NZIM	near-zero index metamaterial
NZI	near-zero index
PRS	partially reflecting surface
RGIMS	radial graded index metasurface
SRR	split-ring resonator

SRRs	split-ring resonators
SRs	spiral resonators
SUT	sample under test
UC	unit cell
UWB	ultra-wideband
VNA	vector network analyser

1

Fundamentals of Metamaterials

1.1 What Are Metamaterials

Metamaterials are artificially engineered homogeneous materials that exhibit unusual electromagnetic (EM) properties not existing in nature. These artificially engineered materials have the capability to modify incoming EM waves in many ways that are beyond naturally existing bulk materials. The Greek word "meta" means beyond, and the term "metamaterial" is used to describe such class of materials which have properties beyond those of existing materials. First proposed in the late 1960s [1], composite materials with engineered optical properties received attention with the promise of perfect lensing. Victor Veselago, the Russian physicist, proposed metamaterials in 1968 [1]. He theoretically investigated the electrodynamics of substances with simultaneously negative values of dielectric permittivity (ε) and magnetic permeability (μ). Veselago also predicted that there exists such a medium in which the electric field, **E**, the magnetic field, **H,** and the wave propagation vector, **k**, form a left-handed orthogonal set [1]. Metamaterials are typically formed by assembling electrically small resonators that take various shapes and compositions [2]. The fact that metamaterials can be engineered to produce an effective medium having simultaneously negative permeability and permittivity has ignited a number of unprecedented applications over various frequency bands. Metamaterials can exhibit negative effective permittivity and negative refractive index, and thus can show refraction or reflection properties unattainable with naturally occurring materials. Such materials can realize effective negative refractive index [3], and therefore, can be utilized to make new devices such as lenses that can image beyond the diffraction limit [2] or invisibility cloaks [4]. Practical realization of such metamaterials is limited due to numerous challenges such as fabrication requirements for layered structures and significant losses in the dispersive materials.

DOI: 10.1201/9781003045885-1

1.2 Unit Cell Concept

The metamaterial's property is due to their structures rather than material composition. These materials are composite, consisting of periodic or non-periodic fundamental sub-wavelength structures. These sub-wavelength structures are known as UCs. These UC structures are similar to the atoms and molecules, controlling the overall EM property of the material. These atoms and molecular-level structures can be engineered using several electric and magnetic resonators. Split-ring resonators (SRRs) [2] are well-known for their magnetic resonance, and wire medium is well-known for its electrical resonance. They have been widely used to design and fabricate metamaterials because SRRs provide the required permeability "μ" negative (MNG) property, whereas wire medium provides the permittivity "ε" negative (ENG) property. SRRs were firstly introduced by Pendry [2], and the first experimental verification of this behaviour was carried out by Smith [3, 4], from the combination of SRRs (magnetic resonators) and metallic rods (electric resonators). However, other magnetic resonators can be used, like spiral resonators (SRs), also initially proposed by Pendry [4], whereas their potential was experimentally assessed by Baena [5]. Additional geometries can also be found in the literature such as capacitive-loaded loops (CLLs) [5] and Ω-particles [6, 7].

The customized EM properties can be generated by optimizing the geometries of the UCs. The cellular architecture of UCs also plays an important role. The EM properties of metamaterials are characterized by their permittivity (ε) and permeability (μ) [8–12]. When the EM structures are effectively homogeneous, these metamaterials behave like real materials and exhibit constitutive parameters: permittivity ε and permeability μ. The refractive index of metamaterial "η" depends on permittivity (ε) and permeability (μ) and can be given by the following equation:

$$\eta = \pm\sqrt{\varepsilon_r \mu_r} \qquad (1.1)$$

Here in Equation 1.1, ε_r and μ_r are the relative permittivity and permeability, respectively. There are four possible combinations of the signs of ε and μ. The double-positive (DPS) materials have both $\varepsilon > 0$ and $\mu > 0$. The epsilon negative (ENG) materials have $\varepsilon < 0$ and $\mu > 0$. The mu negative (MNG) materials have $\varepsilon > 0$ and $\mu < 0$. The double-negative (DNG) material has both $\varepsilon < 0$ and $\mu < 0$ [1, 4]. The plane wave propagation in a material whose permittivity and permeability are negative results in reversed direction of the wave propagation vector **k**, resulting in antiparallel phase and group velocity [1]. These materials are called metamaterials. The incident EM waves on the metamaterial surface interacts with UCs and induces electric and magnetic moments. Due to the induced moments, transmission and reflection characteristics of the material change.

1.3 Metasurface

The arrangement of UC in a plane is known as the metasurface (MS). An MS or a metafilm can be described as a surface that can manipulate the EM waves impinging on them. The manipulation of EM waves can be controlled by the subatomic structures on the top of MS. MSs are thin metamaterial layers characterized by unusual reflection and transmission properties of plane waves and/or dispersion properties of surface/guided waves. MSs can be formed by periodic or aperiodic arrangements of many small inclusions (UCs) in a dielectric host environment, for achieving macroscopic EM or optical properties that cannot be found in nature. The MS is referred to as uniform when the elements in the periodic lattice are identical all over the MS plane. The MS can be designed with aperiodic elements having geometrical parameters gradually changing from cell-to-cell; in this case, the MS is referred to as nonuniform or modulated [13–18]. Nonuniform MS allows one to change the phase velocity and/or propagation path of the guided wave sustained by the MS. These MS structures are also called homogeneous structures. An effectively homogeneous structure is a structure whose structural average cell size is much smaller than the guided wavelength. This average cell size should be at least smaller than a quarter of wavelength. This condition is also called effective homogeneity limit or effective homogeneity condition [18–22]. This condition ensures that inside the structure, refraction phenomena dominate the scattering/diffraction phenomena when an EM wave propagates through the metamaterial medium. If this effective homogeneity condition is satisfied, then the metamaterial structures behave as real materials and myopic to the lattice structure and the structure is called electromagnetically uniform along the direction of propagation of an EM wave. Under this condition, the uniform homogeneous material has some constructive parameters which depend on the nature of the UC. These constructive parameters are permittivity "ε" and permeability "μ". These parameters can be modelled by equivalent electric and magnetic resonance obtained by respective resonators.

1.4 Backward Wave Propagation and Negative Refraction

Metamaterials have unique properties such as backward wave propagation and negative refraction. These properties are due to the material parameters such as permittivity and permeability. The negative values of these parameters cause significant changes in the properties of propagating EM waves. The negative values of material parameters permittivity and permeability cause negative refractive index. Due to this negative refractive index, the phase of wave decreases with propagation in the medium at the place of advancement. This decrease in the phase of the propagating EM wave causes reversal, which in turn results in various important EM phenomena. The reversal of the wave refraction is one of them, which is also called backward wave refraction. As per Snell's law, when an EM wave propagates from one medium to another it bends as per the characteristics of the medium. Snell's law is also the basis of direct measurement of refractive index. The first experiment showing negative refraction was performed in 2001 by Prof. Smith and his group at the University of California, San Diego [4]. They measured the power refracted from a two-dimensional (2D) wedge-shaped metamaterial sample as a function of angle, confirming the expected properties. In 2003, Andrew Houck and colleagues at MIT repeated the negative refraction experiment on the same sort of negative-index metamaterial (NIM) and confirmed the original findings [4, 6]. When the wave vector and group velocity vector of propagating EM waves in an isotropic medium are parallel to each other, they are called forward waves whereas when the wave vector and group velocity vector are antiparallel to each other they are called backward waves. The negative refraction of a parallel wave beam in the layer supporting backward waves leads to a stronger transversal shift of the beam than that in the case of the beam positively refracted in a usual dielectric layer. The energy flux carried by the EM waves can be given by pointing vector "S" that characterizes the surface power density carried by EM waves. When the pointing vector and wave vector are in same direction with E- and H-fields orthogonal to each other, a right-handed triplet formation occurs. These materials are called right-handed materials (conventional materials). Due to existence of backward wave propagation in metamaterials, the pointing vector and the wave vector are in opposite directions and form left-handed arrangements. These types of materials are called left-handed materials (LHMs). In LHMs, the group velocity "V_g" and phase velocity "V_p" are in opposite directions [23–26]. When an EM wave propagates through negative refractive index media, the EM wave bends in the direction of negative angle with respect to the surface normal. Due to this, in such types of media, the concave and convex lenses swap their functionality, i.e. a concave lens will become converging whereas convex lens will become diverging.

Interesting consequences can arise for such a material with negative refractive index. It has been shown that, in such a material, many EM phenomena act in exactly the opposite way to those of conventional materials. When light enters such a material from the outside, it makes a sharp turn, unlike the situation for normal refraction. The Snell's law of "$n_1 \sin\theta_1 = n_2 \sin\theta_2$" still holds, but since n_2 is negative, the refracted ray will remain in the same half-plane (bounded by the normal to the interface) as the incident ray. Similarly, other EM phenomena, such as the Doppler effect and Cherenkov radiation, would exhibit exactly opposite behaviour to that occurring traditionally

[1, 5]. The feature that the wave is a forward or a backward wave is determined only by the inherent medium properties, and these definitions do not refer to refraction phenomena. However, the phenomena of positive/negative refraction, and the forward/backward waves with respect to the interface of the heterogeneous medium are sometimes determined not only by the inherent medium properties, but also by the properties of the interface. Here the orientation of the interface with respect to the inner geometry of the heterogeneous medium plays an important role.

1.5 Split-Ring Resonators

In 1999, it was proposed by Pendry et. al. [2] that creating artificial structures of dimensions smaller than the wavelength of light in a periodic manner can give rise to negative refractive index. These materials were subsequently named as "metamaterials" or LHMs. Pendry proposed that an array of homogeneously spaced sub-wavelength SRRs as metal hoops, which as a whole behave like a composite material. These SRRs can act as an *LC* oscillating circuit containing a magnetic coil of inductance "*L*" and capacitor of capacitance "*C*". In their fundamental resonance, they behave as an *LC* oscillatory circuit that contains a single turn magnetic coil of inductance "*L*" in series with a capacitance "*C*" produced by the gap between the arms of the SRR [6–8].

The dimensions of UC of SRR are very critical and activate differently depending on the polarization of incident EM waves. For normal incidence with transverse electric (TE) polarization mode, when the incident electric field is across the gap (parallel to the *X*-axis), the electric field couples with the capacitance of the SRR and generates a circulating current around it. This circulating current induces a magnetic field in the base of the SRR and also interacts with the external field to generate the magnetic resonance that

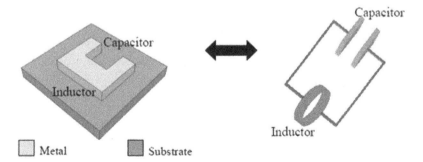

FIGURE 1.1
SRR UC and its equivalent circuit.

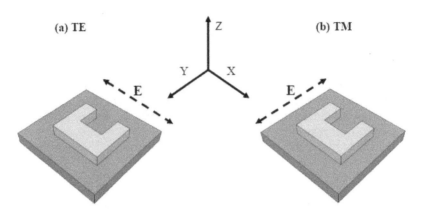

FIGURE 1.2
The SRR with (a) TE-polarized surface incidence having the electric field (dashed line) parallel to the SRR arms (parallel to the X-axis) and (b) TM-polarized surface incidence having the electric field perpendicular to the SRR arms (parallel to Y-axis).

can be appropriately called an *LC* resonanceas identified by Linden et al. [8]. When the incident light is with transverse magnetic (TM) polarization mode, with the electric field perpendicular to the gap (parallel to Y-axis), the electric field cannot couple to the capacitance of the SRR and therefore generates only electric resonances [2–4].

For TE polarization only, the base of the SRR is excited, whereas for TM polarization, both the arms of the SRR are excited, which are coupled via its base. As a result, there is formation of two new symmetric and asymmetric oscillation modes. The asymmetric mode cannot be excited for symmetric SRRs but the symmetric mode gets excited. As described previously, the SRR acts as an *LC* circuit and its magnetic response (*LC* resonance) is due to the circulation of electric current when the electric field of the normally incident wave is parallel to its gap. Hence, the *LC* resonance is polarization-dependent. It is observed that the *LC* resonance wavelength (λ_{LC}) is proportional to the size of the SRR. It is also observed that the gap between arms is very important since if the gap of the SRR is closed, the *LC* resonance frequency becomes zero and the *LC* resonance wavelength (λ_{LC}) becomes infinite [10]. It is also observed that the thickness/height (*h*) of the SRR plays no significant role in determination of its *LC* resonance frequency. Thus, it is seen that the response of an SRR is dependent only on its size, i.e. dimensions and not on metal properties. However, the same is not valid at higher frequencies such as THz. At higher frequencies (optical and near-infrared (IR) regimes), the kinetic energy of the electrons in the metal becomes predominant, and this in turn kills the circulating electric current produced by the coupling of the SRR arms, in turn destroying its *LC* response [11–12]. Due to possible arrangements of SRR, permeability can be controlled and negative permeability can also be generated [27–29]. At the same time, negative permittivity

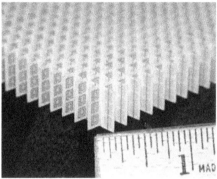

FIGURE 1.3
First experimental metamaterial prototype (a) circular split-ring resonator with cooper wire and (b) rectangular split-ring resonator with transmission line [4].

Source: Copyright/used with permission of/courtesy of AIP.

can be achieved by placing a periodic array of thin metallic wires in close proximity. The wires individually act as electric dipoles and their overall response can give rise to negative permittivity. Combining wires with SRRs can give rise to negative refractive index. The first experimental setup to demonstrate the metamaterial behaviour was designed using this concept.

1.6 Experimental Demonstration of Metamaterial

Prof. Veselago first explained theoretically the existence of metamaterial properties such as negative permittivity, negative permeability, negative refraction and backward wave propagation. However, experimental verifications of such materials with simultaneous negative ε and μ did not occur. In 1996, John B. Pendry proposed that a periodic array of copper wires with a specific radius and spacing produces an EM response of negative ε materials [2]. Later, he further proposed that a periodic array of SRRs would have a frequency band where μ is negative [3].

The first experimental metamaterial prototype is shown in Figure 1.3(a) and (b). David Smith et al. [4] observed that if they could somehow combine these two Pendry's proposed artificial materials [3–4], the composite would be a metamaterial with both negative ε and μ, just as Veselago hypothesized over 30 years ago [1]. This structure was inspired by the pioneering works of Pendry at Imperial College, London. Pendry introduced the plasmonic-type negative-ε/positive-μ and positive-ε/negative-μ structures shown in Figure 1.3(a), which can be designed to have their plasmonic frequency in the microwave range. Both of these structures have an average cell size "p" much smaller than the guided wavelength "λ_g" and are therefore effectively homogeneous structures. The negative-ε/positive-μ metamaterial is the metal thin-wire (TW) structure shown in Figure 1.3(a). When the excitation electric field **E** is parallel to the axis of the wires, it induces a current along the wire and generates equivalent electric dipole moments [8]. This metamaterial exhibits a plasmonic-type permittivity frequency [5, 6]. On the other hand, permeability is simply $\mu = \mu_0$, since no magnetic material is present and no magnetic dipole moment is generated. It should be noted that the wires are assumed to be much longer than wavelength, which means that the wires are excited at frequencies far below their first resonance. The positive-ε/negative-μ metamaterial is the metal split-ring resonator (SRR) structure shown in Figure 1.3 (a) and (b). If the excitation magnetic field **H** is perpendicular to the plane of the rings, it induces resonating currents in the loop which generate equivalent magnetic dipole moments. This metamaterial exhibits a plasmonic-type permeability frequency function. The first designed metamaterial consists of the following: (1) a thin wire structure and an SRR structure with overlapping frequency ranges of negative permittivity and permeability; (2) combining the two structures into a composite thin wire–SRR structure, which is shown in Figure 1.3(a); and (3) launching an EM wave through the structure and recognizing the pass-band characteristics [30].

References

1. Veselago, V. G., "The electrodynamics of substances with simultaneously negative values of ε and μ," *Physics-Uspekhi*, vol. 10, no. 4, pp. 509–514, April 1968.
2. Pendry, J. B., A. J. Holden, D. J. Robbins, and W. J. Stewart, "Magnetism from conductors and enhanced nonlinear phenomena," *IEEE Transactions on Microwave Theory Technique*, vol. 47, no. 11, pp. 2075–2084, Nov. 1999.
3. Pendry, J. B., "Negative refraction makes a perfect lens," *Physical Review Letters*, vol. 85, no. 18, pp. 3966–3969, Oct. 2000.
4. Smith, D. R. D. Schuring, and J. B. Pendry, "Negative refraction of modulated electromagnetic waves," *Applied Physics Letters*, vol. 81, pp. 2713–2715, 2002.
5. Juan Domingo, B., Jordi Bonache, Ferran Martin, R. Marqués Sillero, Francisco Falcone, Txema Lopetegi, Miguel A. G. Laso et al., "Equivalent-circuit models for split-ring resonators and complementary split-ring resonators coupled to planar transmission lines," *IEEE Transactions on Microwave Theory and Techniques*, vol. 53, no. 4, p. 1451–1461, 2005.
6. Valanju, P. M., R. M. Walser, and A. P. Valanju, "Wave refraction in negative-index media: Always positive and very inhomogeneous," *Physical Review Letters*, vol. 88, no. 18, p. 187401, Jan. 4, 2002.
7. Linden, S., C. Enkrich, G. Dolling, M. W. Klein, J. Zhou, T. Koschny, C. M. Soukoulis, S. Burger, F. Schmidt, and M. Wegener, "Photonic metamaterials: Magnetism at optical frequencies," *Journal of Selected Topics in Quantum Electronics*, vol. 12, no. 6, pp. 1097–1103, 2006.

8. Zhou, J., T. Koschny, M. Kafesaki, E. N. Economu, J. B. Pendry, and C. M. Soukoulis, "Saturation of the magnetic response of split ring resonators at optical frequencies," *Physics Review Letters*, vol. 95, no. 223902, pp. 1–4, 2005.
9. Klein, M. W., C. Enkrich, M. Wegener, C. M. Soukoulis, and S. Linden, "Single split ring resonators at optical frequencies: Limits of size scaling," *Optics Letters*, vol. 31, pp. 1259–1261, 2006.
10. Gay-Balmaz, P. and O. J. F. Martin, "Electromagnetic resonances in individual and coupled split-ring resonators," *Journal of Applied Physics*, vol. 92, no. 5, pp. 2929–2936, Sept. 2002.
11. Marqués, R., F. Median, and R. Rafii-El-Idrissi, "Role of bianisotropy in negative permeability and left-handed metamaterials," *Physical Review B*, vol. 65, pp. 144440, Jan.6, 2002.
12. Markös, P. and C. M. Soukoulis, "Numerical studies of left-handed materials and arrays of split ring resonators," *Physical Review E*, vol. 65, pp. 036622, Jan.8, 2002.
13. Greegor, R. B., C. G. Parazzoli, K. Li, B. E. C. Koltenbah, and M. Tanielian, "Experimental determination and numerical simulation of the properties of negative index of refraction materials," *Optics Express*, vol. 11, no. 7, pp. 688–695, April 2003.
14. Ozbay, E., K. Aydin, E. Cubukcu, and M. Bayindir, "Transmission and reflection properties of composite double negative metamaterials in free space," *IEEE Transactions on Antennas and Propagation*, vol. 51, no. 10, pp. 2592–2595, Oct. 2003.
15. Simovski, C. R., P. A. Belov, and S. He, "Backward wave region and negative material parameters of a structure formed by lattices of wires and split-ring resonators," *IEEE Transactions on Antennas and Propagation*, vol. 51, no. 10, pp. 2582–2591, Oct. 2003.
16. Marqués, R., F. Mesa, J. Martel, and F. Median, "Comparative analysis of edge- and broadside-coupled split ring resonators for metamaterial: Design, theory and experiments," *IEEE Transactions on Antennas and Propagation*, vol. 51, no. 10, pp. 2572–2581, Oct. 2003.
17. Lindell, V., S. A. Tretyakov, K. I. Nikoskinen, and S. Ilvonen, "BW media–media with negative parameters, capable of supporting backward waves," *Microwave and Optical Technology Letters*, vol. 31, no. 2, pp. 129–133, Oct. 2001.
18. Kong, J. A., B.-I. Wu, and Y. Zhang, "A unique lateral displacement of a gaussian beam transmitted through a slab with negative permittivity and permeability," *Microwave and Optical Technology Letters*, vol. 33, no. 2, pp. 136–139, April 2002.
19. Pacheco, J., T. M. Grzegorzcyk, B.-I. Wu, Y. Zhang, and J. A. Kong, "Wave propagation in homogeneous isotropic frequency-dispersive left-handed media," *Physical Review Letters,* vol. 89, no. 25, p. 257401, Dec. 4, 2002.
20. Smith, D. R., D. Schurig, and J. B. Pendry, "Negative refraction of modulated electromagnetic waves," *Applied Physics Letters*, vol. 81, no. 15, pp. 2713–2715, Oct. 2002.
21. McCall, M. W., A. Lakhtakia, and W. S. Weiglhofer, "The negative index of refraction demystified," *European Journal of Physics*, vol. 23, pp. 353–359, 2002.
22. Iyer, K. and G. V. Eleftheriades, "Negative refractive index metamaterials supporting 2-D waves," *IEEE-MTT International Symposium*, vol. 2, Seattle, WA, pp. 412–415, June 2002.

23. Eleftheriades, G. V., A. K. Iyer, and P. C. Kremer, "Planar negative refractive index media using periodically L-C loaded transmission lines," *IEEE Transactions on Microwave Theory and Techniques*, vol. 50, no. 12, pp. 2702–2712, Dec. 2002.
24. Caloz, C. and T. Itoh, "Application of the transmission line theory of left-handed (LH) materials to the realization of a microstrip LH transmission line," *Proceedings of the IEEE-AP-S USNC/URSI National Radio Science Meeting*, vol. 2, San Antonio, TX, pp. 412–415, June 2002.
25. Belov, P. A., "Backward waves and negative refraction in uniaxial dielectrics with negative dielectric permittivity along the anisotropy axis," *Microwave and Optical Technology Letters*, vol. 37, no. 4, pp. 259–263, March 2003.
26. Felbacq, D. and A. Moreau, "Direct evidence of negative refraction media with negative ε and μ," *Journal of Optics A*, vol. 5, pp. L9–L11, 2003.
27. Lakhtakia, "Positive and negative Goos-H anchen shifts and negative phase-velocity mediums (alias left-handed materials)," *International Journal of Electronics and Communications*, vol. 58, no. 3, pp. 229–231, 2004.
28. Al'u and N. Engheta, "Guided modes in a waveguide filled with a pair of single negative (SNG), double-negative (DNG) and/or double-positive (DPS) layers," *IEEE Transactions on Microwave Theory and Techniques*, vol. 52, no. 1, pp. 192–210, Jan. 2004.
29. Grbic, A. and G. V. Eleftheriades, "A backward-wave antenna based on negative refractive index L-C networks," *Proceedings IEEE-AP-S USNC/URSI National Radio Science Meeting*, vol. 4, San Antonio, TX, pp. 340–343, June 2002.
30. Caloz, C., H. Okabe, T. Iwai, and T. Itoh, "Anisotropic PBG surface and its transmission line model," *URSI Digest, IEEE-AP-S USNC/URSI National Radio Science Meeting*, San Antonio, TX, p. 224, June 2002.

2

Design, Fabrication and Testing of Metamaterials

2.1 Design of Metamaterials

The EM waves propagating through a medium interact with it at atomic or molecular level to generate the resultant wave. The material properties of the medium can control the properties of the propagating wave through it. Using the material properties such as permittivity "ε" and permeability "μ", the medium interacts with electric field and magnetic field of the propagating EM waves. Permittivity is the property of medium to permit flow of electric field through the medium and permeability is the property of medium to permit flow of magnetic field through the medium. The permittivity of the medium can be controlled by controlling the capacitive behaviour of the medium and permeability can be controlled by controlling the inductive behaviour of the medium. This controlled capacitive and inductive behaviour of the medium can be achieved if the source of capacitive behaviour of the medium and source of inductive behaviour of the medium can be tailored at subatomic level. The controlled permittivity and permeability can generate artificial medium having desired positive or negative values of permittivity "ε" and permeability "μ".

Microstrip structures having optimum dimensions can generate proper values of inductance and capacitance, resulting in a medium having controlled material parameters like permittivity "ε" and permeability "μ". The optimum dimensions of microstrip structures can generate negative values of permittivity "ε" and permeability "μ". Achieving negative permeability is more difficult as compared to achieving negative permittivity [1–6]. To achieve negative permittivity using microstrip structures, the source of capacitance, such as microstrip gaps, slits and slots in between the metal conducting structures, can be optimally utilized. The simulated electric field distribution using full-wave EM simulators due to normal or

DOI: 10.1201/9781003045885-2

in-plane incidence of EM waves can be analysed to find out the source of the capacitive behaviour. Highly concentrated electric field regions on the surface of the microstrip structure can be one of the possible sources of capacitive behaviour of the microstrip geometry or structure. The optimized dimensions of such geometries can result in optimum desired values of permittivity as per the requirement. Hence, controlled permittivity characteristic can be achieved. Similarly, to achieve negative permeability using microstrip structures, the source of inductance can be optimized. The simulated magnetic field distribution using a full-wave simulator due to normal or in-plane incidence of EM waves can be analysed to find out the source of inductive behaviour. Highly concentrated magnetic field regions on the surface of the microstrip structure can be one of the possible sources of the inductive behaviour [7–11]. Optimized dimensions of such geometries can result in desired permeability characteristic. However, achieving desired permeability characteristic is a bit difficult as compared to desired permittivity.

The artificial media, when designed in such a way that the electrical dimensions are much smaller than the propagating wavelength in the microstrip structures, can be called sub-wavelength structures. The periodic or non-periodic arrangement of these sub-wavelength structures can result in a medium having controllable permittivity and permeability. These sub-wavelength structures are called UCs or meta-atoms, and the designed artificial dielectric material is called metamaterial. Metamaterial designed in a planar 2D surface is known as MS. The architecture of UCs depends on two parameters, the operating frequency and the angle of incident wave on the structure. Incident EM waves on the UC parallel to UC axis induce an electric field on the inductive region of the microstrip UC geometry. Similarly, it induces magnetic field on the capacitive region of microstrip UC geometry. The induced electric and magnetic fields generate resonance, where resonant frequency can be determined by equivalent inductance and equivalent capacitance of the UC structure. This resonant frequency due to induced electric and magnetic field is called resonance frequency of UC or meta-atom. Depending on the geometries of the UC, a UC may have more than one resonant frequency, wideband width resonant frequency and transmission- or reflection-type resonant characteristic. The fundamental UC (FUC) geometry can be rectangular, circular, elliptical or any irregular geometry. It is observed that the outer dimension of the resonating UC is equal to half resonating wavelength. The UC dimensions can be made more compact using capacitive and inductive loading. As the UC and MS are planar, it can be fabricated using photolithography or three-dimensional (3D) printing. The MS can be of resonant or non-resonant type. The UC structures determine their transmission and reflection band characteristics.

2.2 Characterization of Metamaterial and Measurement Techniques

The designed and fabricated metamaterial surfaces (2D or 3D) can be tested for their transmission and reflection characteristics using several techniques. The MS can be excited using the normal or oblique incidence of EM waves on the surface. The incident EM waves on the surface induces surface currents, which results in various modes of induced EM waves on the surface. Various combinations of inductive and capacitive characteristics of the UC of MS result in different operations [8, 9]. Depending on the operational behaviour of UC/meta-atom of MS, the incident EM waves can be reflected or transmitted. To characterize the MS, two methods of characterization, resonant and non-resonant methods, can be used. Out of these two, the resonant method of characterization is more accurate. However, the resonant method has limitations; the method is suitable only for narrow band or for a single frequency. The non-resonant method is suitable for a wide frequency range. Non-resonant methods are also known as the reflection methods and the transmission/reflection methods. In non-resonant methods, the partially reflected and transmitted EM waves from the surface are analysed to find out the impedance and wave velocity characteristics of the material under test followed by extraction of EM characteristics such as permittivity and permeability.

In non-resonant methods, a wave director is used to direct the wave to material under test and then to collect the transmitted and/or reflected EM waves. This wave director can be a waveguide (WG), planar transmission line, a microwave radiator and free space [12–16]. The transmitted and/or reflected EM waves can be recorded and measured using wave directors and a vector network analyser (VNA). During measurement using VNA, three types of errors can occur: systematic error, random error and drift error [17–22]. Systematic error is mainly due to directivity, cross-talk, multiple reflections and frequency response of the wave director and the material under test. Random errors are unpredictable in nature, and they can be due to many unpredictable conditions during measurement, and the same can be removed by making several measurements and taking the average values [16]. Drift errors are mainly due to change in working conditions and require a stable environment in terms of temperature and humidity.

The aforementioned errors can be removed using many techniques. Systematic errors can be removed using calibration techniques. In the process of calibration, systematic errors due to measurement are calculated using several known standards, and the same can be removed mathematically by taking many subsequent measurements. Two types of error corrections can be done—response correction and vector correction. Vector correction is the most important one and can account for all the major sources of systematic errors using magnitude and phase data of the network analyser. Similarly, by doing additional calibration, the drift and random errors can also be removed.

2.2.1 Non-Resonant Methods of Metamaterial Characterization

The characterization of MSs for wideband operation can be done by non-resonant methods. There are two types of non-resonant methods—reflection method and transmission/reflection method. In the reflection method, a transmission line or wave director is loaded with the metamaterial surface under test at a certain position. The placement of the material under test at a certain position causes impedance loading. This discontinuity in impedance can be used to derive the relative permittivity and relative permeability [15]. There are two types of reflection measurement methods, the open-circuit reflection method and short-circuit reflection method. The permittivity and permeability can be calculated if sufficient numbers of independent reflection measurements are performed. However, in most cases, only one independent measurement is made, so only one material property can be obtained. To obtain both permittivity and permeability, more complex experimental set-ups, such as probes and sensors with transmission line or wave director, are needed. In the open-circuited reflection method, the MS under test directly contacts the open end of the coaxial line, probe or wave director. The imped-ances at the two sides of the interface are different, and there is reflection when the EM wave is incident to the interface. The reflectivity is determined by the impedances of the media at the two sides of the interface. The imped-ance of the sample's side is related to the properties of the sample, and from the reflectivity at the interface, the properties of the sample can be obtained.

Next type of non-resonant method of MS characterization is transmission/reflection method. In this measurement method, the sample under test (SUT) is placed in a segment of transmission line, usually a WG or a wave director such as an antenna. The EM properties of the material under test, permittiv-ity and permeability, can be derived from the measured scattering param-eters. The reflected and transmitted EM waves from the MS under test can be recorded using VNA. The collected scattering parameters can be post-processed using several algorithms, and the material properties such as per-mittivity and permeability, can be extracted. In such a method, all the four scattering parameters need to be measured. The complex permittivity and permeability, the sample length and the positions of the two reference planes are the main variables contained in the relevant scattering parameter equa-tions of a transmission/reflection measurement. Although for a coaxial line the cut-off wavelength is infinity, the more difficult placement of a sample in such a line leads to more often use of WGs, especially rectangular ones. The proper size of the sample as well as proper placement of the sample in the far-field region of the microwave radiator is a necessary condition in this type of measurement technique. Once the scattering parameters are recorded, the same data can be post-processed and material property such as permittivity and permeability can be extracted. One of the most commonly used methods for material parameter extraction is Nicolson–Ross–Weir (NRW) algorithm [2].

2.2.2 Nicolson–Ross–Weir (NRW) for Parameter Extraction

In the NRW algorithm, it is assumed that the SUT is excited with the normal incidence of EM waves [2]. The scattering parameter data can be recorded under this condition using VNA and EM wave radiator. There are several major difficulties in implementing the NRW algorithm, one being the sample size that should be near a multiple of half-wavelength. To analyse the scattering parameters using NRW algorithm, composite terms such as "V_1" and "V_2" are introduced.

$$V_1 = S_{21} + S_{11}$$
$$V_2 = S_{21} - S_{11}$$

(2.1)

Similarly, defining the terms "X and Y"

$$X = \frac{1 + V_1 V_2}{V_1 + V_2} = \frac{1 + Z^2}{2Z}$$
$$Y = \frac{1 - V_1 V_2}{V_1 - V_2} = \frac{1 + \Gamma^2}{2\Gamma}$$

(2.2)

Similarly, from Equation 2.2, we can obtain the values of "Z and Γ", transmission and reflection terms, respectively, as

$$Z = X \pm \sqrt{X^2 - 1}$$
$$\Gamma = Y \pm \sqrt{Y^2 - 1}$$

(2.3)

Many other expressions of "Z and Γ" can also be derived and the same are given as

$$Z = \frac{V_1 - \Gamma}{1 - \Gamma V_1}$$

(2.4)

$$\Gamma = \frac{Z - V_2}{1 - Z V_2}$$

(2.5)

From Equations 2.4 and 2.5, we can obtain a direct expression as follows:

$$1 - Z \quad = \frac{(1 - V_1)(1 + \Gamma)}{1 - \Gamma V_1}$$

$$\eta \quad = \frac{1 + \Gamma}{1 - \Gamma} = \frac{1 + Z}{1 - Z} \frac{1 - V_2}{1 + V_2} \,.$$ (2.6)

Assuming the thickness of the SUT to be very small and considering several approximations, the permittivity and index of refraction can then be obtained simply as follows:

$$\epsilon_r \quad = \left(\frac{k}{k_0}\right)^2 \frac{1}{\mu_r}$$

$$n \quad = \sqrt{\varepsilon_r \mu_r} = \frac{k}{k_0}$$ (2.7)

Here, the complex wave number $k = \omega \sqrt{\varepsilon_r \mu_r} / c = k_0 \sqrt{\varepsilon_r \mu_r}$, and using several approximations, the transmission coefficient can be written as $Z \sim 1 - jkd$. Similarly, "ε_r" can be calculated as

$$\varepsilon_r \sim \frac{2}{jk_0 d} \frac{1 - V_1}{1 + V_1} \,.$$ (2.8)

Also, Equation 2.8 can be rewritten as

$$\varepsilon_r \approx \mu_r + j \frac{2S_{11}}{k_0 d} \,.$$ (2.9)

Equation 2.9 explicitly shows that ε_r and μ_r should exhibit very similar responses when S_{11} is approximately zero. Applying the modified NRW technique using Equations 2.7, 2.8 and 2.9, the material parameters of metamaterial surface under test can be extracted. Due to the various characteristics of metamaterials, most of the time at resonance conditions, the values of transmission and reflection terms become zero and unity. Hence, the standard extraction techniques and expressions are not found suitable for parameter extraction, particularly in those frequency bands where resonance is expected [2], i.e. where the permittivity and permeability follow quick transition between positive and negative values.

There are several other techniques available for non-resonant transmission and reflection methods for metamaterial characterization. Out of these WG method and free space method are more accurate and popular among the reported techniques. In both measurement methods, a plane EM wave incidence on the material under test is required to excite the metamaterial surface. The reflection and transmission characteristics can be recorded using VNA, and material parameters can be extracted with the equations mentioned earlier using modified NRW techniques [6].

2.2.3 Waveguide Measurement Technique

In WG measurement method, the sample material under test is machined to fit inside a WG sample holder in direct contact with WG adaptors. This measurement technique consists of two WG ports, a sample holder, a WG transition and WG ports connected to the VNA [22–26]. Due to WGs, this measurement technique limits the frequency range of measurement. A VNA is connected to both WG adaptors to measure transmission and reflection properties of the sample. The sample holder with WG transitions and WG ports forms a tightly coupled cavity. The WG measurement setup is shown in Figure 2.1. Any kind of wave leakage from the cavity or sample holder results in improper measurement of transmission and reflection characteristics.

However, when a rectangular WG filled with a sample is to be analysed, the calibration reference plane for measuring scattering parameters needs to be transformed. The WG consists of two ports and WG transition connected with a sample holder between them. The sample is placed in the sample holder, and scattering characteristics are recorded by the VNA. However, due to initial level of calibration procedure inside the VNA, the calibration planes are set at the end of the WG ports only and scattering parameters are recorded at these ports. To determine the scattering parameters of the actual sample inside the holder accurately, we use a technique called "shift in reference plane" to get scattering parameters at the sample edges inside the WG. This transformation is very important especially when the phase of the scattering parameters is to be studied. In the case of an air gap existence between the sample and the WG, the measured EM properties (the output of

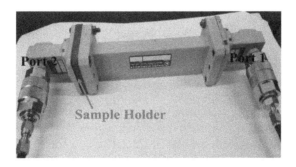

FIGURE 2.1
WG measurement setup for material characterization

the NRW algorithm) also need to be modified with gap properties, permittivity and permeability of the gap medium. Any kind of improper calibration or measurement error may cause error in the extracted parameter values as well.

2.2.4 Free Space Measurement Technique

Free space measurement technique is contactless and non-destructive; hence, it is more suitable for complex permittivity and complex permeability measurements. This method is more suitable for broadband measurement of material properties of high-loss materials. In free space method, the sample material under test does not make any direct contact with radiating devices used in the measurement setup. In this measurement technique, the free space transmission coefficient and reflection coefficient of a planar sample due to normal incidence of EM waves are measured [26]. The measurement system consists of two wideband horn antenna radiators, a sample holder placed in the far-field region of both the radiators, a VNA, mode transitions and a computer. The errors in the measurements are found to be due to diffraction effect at the edge of the sample and multiple reflections between horn lens antennas and the mode transitions. To avoid the effect of diffraction at the edges of the sample, short-focusing lens antenna can be designed. The short-focusing horn lens antenna consists of two equal metamaterial lenses or two equal plano-convex dielectric lenses mounted back to back in a conical horn antenna. In addition to this, a free space calibration technique TRL (thru, reflect, line) with time domain gating is used to cancel out the effect of multiple reflections and mode transition on the surface of the sample [27–30]. This TRL calibration can be verified for plane wave propagation by measuring free space "S_{21}" for free space delays of different lengths. A linear phase variation and negligible magnitude variation can be observed for stable calibration. A specially fabricated sample holder can be designed and placed in the common far-field region or at the common focal plane.

The SUT is placed using a sample holder in the far field of the radiator. Two standard gain wideband radiators, acting as a transmitter and receiver, connected to a network analyser, as shown in Figure 2.2, are used. The transmission and reflection characteristics of sample material are recorded using VNA. For measuring the transmission and reflection characteristics of the sample material, the reference planes corresponding to transmit and receive antennas are relocated to front and back faces of the sample, respectively. Time gating is applied to record the transmission and reflection characteristics. The material parameters are extracted using measured transmission and reflection characteristics with the help of standard extraction methods like NRW method as in [22] and as discussed in Section 2.3.1. Two modes of free space method can be used, transmission-mode free space measurement and reflection-mode free space measurement.

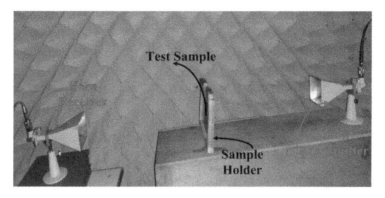

FIGURE 2.2
Free space measurement setup in anechoic chamber.

Reflection mode free space measurement technique is one-port measurement consisting of one wideband horn radiator connected to the VNA, and the SUT is backed with uniform conductor sandwiched and placed in the far field of the horn antenna radiator. The reflection coefficient is measured under normal incidence of the EM wave. The reflection characteristics can be used to extract material parameters. The transmission line theory is used to derive the complex permittivity of the material [22]. This method is not suitable for magnetic material characterization.

Transmission-mode free space measurement technique is a two-port measurement consisting of two wideband horn radiators, acting as transmitter and receiver, connected to the VNA and the SUT is placed inside a sample holder in between both radiators. The measurement setup is shown in Figure 2.2. The reflected and transmitted energy through the test sample is measured and used for extracting complex permittivity and permeability. The standard NRW algorithm is used to extract the material parameters such as complex permittivity and permeability. This method is modified by Ziolkowski in [23] as reported in Section 2.3.1. The same method is used for material parameter extraction in this book. This method is suitable for all types of material characterizations. This technique is contactless, gives results over wideband and is a more efficient material characterization method. But the extracted material properties depend on sample thickness; a relatively larger sample size is required for testing the sample.

A miniaturized free space measurement technique is designed and used for material characterization in this work also. The measurement setup consists of two standard WG port radiators acting as the transmitter and the receiver. To reduce the error due to diffraction, the focus of the WG port radiator can be reduced by using metamaterial lenses. The test sample is placed in the far-field region of the transmitter and the receiver to avoid

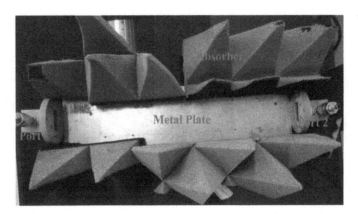

FIGURE 2.3
Miniaturized free space measurement setup.

interference. The microwave pyramidal absorbers are applied on the two sides of the measurement setup to minimize the extra interference. The absorption level of absorber used is seen to be less than −30 dB. The complete setup is designed on a thick conducting plate to realize approximately plane wave incidence on the test sample. The measurement setup is shown in Figure 2.3. This measurement setup is the modified form of the measurement setup used in [23]. The free space measurement without material under test is done first and a reference is established. Then the designed metamaterial is inserted in the sample holder and scattering parameters are captured using VNA and time gating technique. The designed measurement setup with modified NRW algorithm, as reported in [22] and [23], is used to extract the material parameters, permittivity and permeability, and characterize the MS.

The normal incident plane wave on the test sample suffers from two major reflections during complete propagation. The first reflection occurs on the front surface due to impedance mismatch between free space and the test sample. Second reflection occurs on the backside surface due to impedance mismatch between test sample and the free space [28–33]. When the test sample thickness is an integer multiple of half propagating wavelength in the medium then the second reflected wave interferes destructively with the forward-propagating wave inside the sample. This destructive interference dominates the propagating dielectric loss if the losstangent of the material is small. This effect causes a significant error in the NRW method for material parameter extraction algorithm. To avoid this error, TRL calibration technique at VNA and sample thickness restriction is followed. In the studies reported in this book, we have avoided the aforementioned conditions [27]. The thickness of MS designed in our work is much smaller than propagating wavelength. This effect is observed only in transmission-mode free space measurements.

2.3 Design of a Wideband Low-Profile Ultrathin Metasurface for X-Band Application

In this section, design of a low-profile ultrathin miniaturized wideband reflection-type metamaterial surface is discussed. The bandwidth enhancement technique and electrical miniaturization techniques are also reported. The measured 3-dB stop-band bandwidth of 3.98 GHz from 8.02 GHz to 12.00 GHz is achieved using this approach. The prototype MS is fabricated and measured. Overall bandwidth enhancement of 50 per cent with electrical size reduction of 53.64 per cent is achieved [34].

2.3.1 Design of the Metasurface and Experimental Characterization

To design a wideband reflection-type MS, initially a circular ring resonator (CRR) is considered. The outer mean circumference dimension is selected as one half of resonating wavelength. The width of the metallic section ring is optimized to get the desired wideband. The CRR UC is designed on Neltec substrate with permittivity 2.2 and thickness of 0.762 mm. The CRR UC with detailed dimensions is shown in Figure 2.4(a). The UC is simulated using CST microwave studio software with perfect electric conductor (PEC) and perfect magnetic conductor (PMC) boundaries with two WG ports to ensure normal incidence of EM waves on the CRR as shown in Figure 2.4(b).

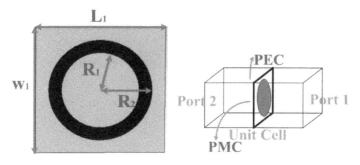

FIGURE 2.4

(a) CRR UC and (b) simulation boundary condition for any UC sample. Dimensions are $L_1 = 14$, $W_1 = 14$, $R_1 = 5$, $R_2 = 7$ (all dimensions are in millimetres).

Source: Copyright/used with permission of/courtesy of IEEE.

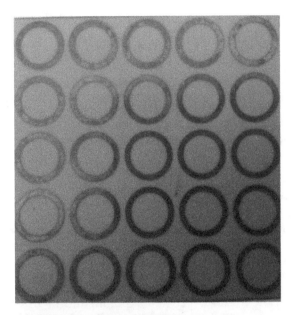

FIGURE 2.5
(a) The fabricated CRR metasurface and (b) measurement setup.

Source: Copyright/used with permission of/courtesy of IEEE.

A 5 × 5 array of CRR UC is designed and fabricated using photolithography on Neltec substrate with permittivity 2.2 and thickness 0.762 mm. The fabricated MS is shown in Figure 2.5(a). The fabricated CRR UC MS is measured using free space measurement setup reported in this section. The measurement setup with two X-band WG port radiators and absorbers attached at both sides is designed on metal surface to generate plane wave. The measurement setup is shown in Figure 2.5(b).

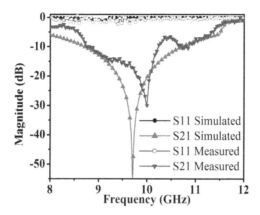

FIGURE 2.6
Measured and simulated transmission and reflection characteristics.

Source: Copyright/used with permission of/courtesy of IEEE.

The measured and simulated reflection and transmission characteristics of 5 × 5 array of CRR UC MS are compared in Figure 2.6. A good agreement between measured and simulated results is observed. Small mismatch in minimum measured transmission characteristic observed is due to alignment error. The improper alignment of WG port generates improper modes on MS, resulting in slight error. Measured 3-dB stop bandwidth of about 2.0 GHz from 9.0 GHz to 11.0 GHz is obtained. The material parameters of the fabricated MS are extracted from measured transmission and reflection characteristics using free space technique.

To get wideband reflection characteristics with miniaturization of UC, the CRR UC is modified and a modified circular SRR (MCsRR) is designed. The MCsRR is a modified form of CRR where a rectangular split is created. It is observed that the surface current concentration is maximum on the split gap sides, resulting in a capacitive effect. To further miniaturize the UC, the capacitive split is modified to introduce more capacitive effect, resulting in miniaturization of the CRR UC. The MCsRR UC with detailed dimensions is shown in Figure 2.7(a). The simulation boundary conditions are shown in Figure 2.7(b). The slot penetration depth "h" is a very important parameter that controls accumulation of surface current due to which centre stop-band frequency can be decreased due to capacitive slot loading and more conductor thickness. This technique provides improvement in quality factor of the resonator, resulting in an improvement in bandwidth. The material parameters are extracted using free space method as shown in Section 2.3.1. The extracted material parameter of MCsRR UC MS is shown in Figure 2.8(a). A 10 × 10 array of MCsRR UC is fabricated on Neltec substrate and the fabricated prototype is shown in Figure 2.8(b).

The transmission and reflection characteristics of designed MCsRR UC MS are measured using free space measurement setup as shown in Figure 2.5

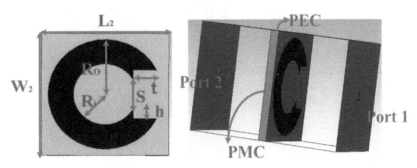

FIGURE 2.7

(a) MCsRR UC geometry and (b) simulation boundary condition. Dimensions are $L_2 = 7$, $W_2 = 7$, $R_o = 3.5$, $R_1 = 2$, $S = 1.75$, $h = 0.5$ and $t = 1.25$ (all dimensions are in millimetres).

Source: Copyright/used with permission of/courtesy of IEEE.

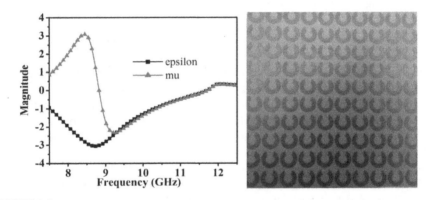

FIGURE 2.8

(a) Extracted material parameter of MCsRR UC MS and (b) fabricated MCsRR UC MS.

Source: Copyright/used with permission of/courtesy of IEEE.

(b). The fabricated prototype is placed in between WG port radiators to avoid near-field effect. The simulated and measured reflection and transmission characteristics of MCsRR UC MS are compared in Figure 2.9. A good agreement between measured and simulated results is observed. A small mismatch in measured transmission dip is observed. This is due to alignment problems and finite size of MS for measurement. A wide reflection band from 8.02 GHz to more than 12.0 GHz with 3-dB stop bandwidth of more than3.98 GHz is observed. The electrical size of MCsRR UC is miniaturized by 53.64 per cent with 3-dB stop bandwidth enhancement of about 50 per cent as compared to electrical size and 3-dB stop bandwidth of CRR UC MS.

In summary, a wide reflection band with low-profile ultrathin MCsRR UC MS is designed and characterized successfully. A 3-dB stop bandwidth enhancement of 50 per cent with miniaturization of about 53.64 per cent as compared to CRR UC MS is obtained. The electrical size of MCsRR UC is

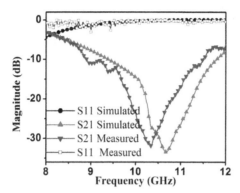

FIGURE 2.9
Simulated and measured transmission and reflection characteristics of MCsRR UC MS.

Source: Copyright/used with permission of/courtesy of IEEE.

$0.25\lambda \times 0.25\lambda \times 0.025\lambda$, where λ is wavelength at reflection band centre frequency of 10.30 GHz. The proposed MS is ultrathin and low-profile with miniaturization.

2.4 Design of a Low-Profile Ultrathin Multiband Transmission- and Reflection-Type Metasurface

In this section, design of a multiband transmission- and reflection-type MS for C-band, X-band and Ka-band applications is reported. A single MS with multiple bands of operation with transmission and reflection characteristics is designed. The designed MS is found to be ultrathin and compact. The prototype of proposed MS is fabricated and tested. The MS is experimentally characterized for its transmission and reflection characteristics [35, 36].

2.4.1 Design of the Metasurface

The UC is designed by considering a square patch as reference and then modifying it in multiple steps. The proposed metamaterial UC is a double annular slot ring resonator (DASR). The design steps of DASR are shown in Figure 2.10 as (a), (b), (c), (d), (e) and (f), and their transmission and reflection characteristics are shown in Figure 2.11. Loading of circular slotted square patch (a) with annular ring (b) generates one reflection band in between two transmission bands (Figure 2.11). Further, loading of annular ring with concentric circular disc (c) causes shifting of all resonant frequencies to the lower side due to capacitive loading. Loading of annular ring with metallic serrations (d) causes small shift of resonant frequencies with enhanced 3-dB bandwidth. Further loading of serrated annular ring with concentric circular disc (e) causes very small shift in first and second resonant frequencies while

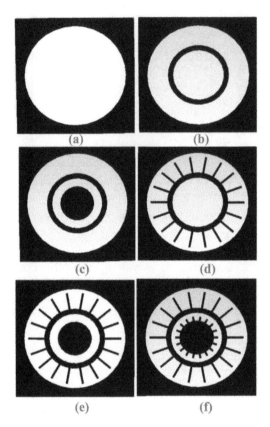

FIGURE 2.10
Design steps of proposed DASR UC.

Source: Copyright/used with permission of/courtesy of IEEE.

third resonant band is disturbed. Loading of structure (e) with metallic serrations on concentric circular disc (f) causes lower-side shifting of resonant bands with improved transmission band at the third resonant frequency (Figure 2.11). The increasing number of serrations increases capacitive loading, resulting in decrease of resonant frequency and vice versa. Hence, the proposed DASR metamaterial UC is novel in terms of compactness and wide 3-dB stop-band bandwidth.

The detailed geometry of proposed DASR UC with dimensions is shown in Figure 2.12(a). The proposed UC is simulated using CST microwave studio with PEC and PMC boundary conditions as shown in Figure 2.12(b).

Simulated transmission and reflection parameters of the UC are shown in Figure 2.13. The simulated characteristics show that the proposed structure is resonating at 10.20-GHz frequency with wide stop-band characteristics from 7.50 GHz to 13.25 GHz. The proposed structure has two more transmission bands with centre frequency 6.25 GHz and 15.30 GHz, respectively.

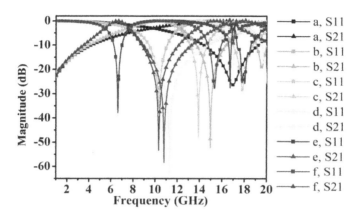

FIGURE 2.11
Simulated transmission and reflection characteristics of design steps in Figure 2.10.

Source: Copyright/used with permission of/courtesy of IEEE.

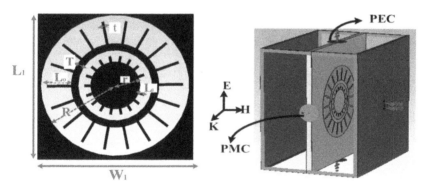

FIGURE 2.12

(a) DASR UC with dimensions and (b) simulation boundary condition. Dimensions are $L_1 = 13$, $W_1 = 13$, $R = 6$, $r = 2$, $L_i = 0.6$, $L_e = 2.04$ and $T = t = 0.2$ (all dimensions are in millimetres).

Source: Copyright/used with permission of/courtesy of IEEE.

The simulation results indicate that this UC can be used for multiple applications at a time. A planar MS is designed using 5×5 array of proposed novel DASR UC, as shown in Figure 2.14(a). The material property of the proposed structure is extracted from simulated transmission and reflection parameters using free space technique as reported in [37] and in Section 2.3.1. The extracted material permittivity and permeability as function of frequency are plotted in Figure 2.14(b). The proposed structure is showing negative permeability and permittivity in the desired band of operation.

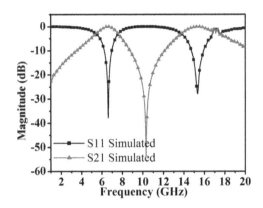

FIGURE 2.13

Simulated transmission and return loss parameter of UC.

Source: Copyright/used with permission of/courtesy of IEEE.

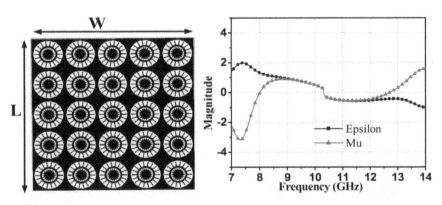

FIGURE 2.14

(a) Proposed novel DASR meta-surface and (b) extracted material parameter of metamaterial UC. Dimensions are (L = 65 mm and W = 65 mm).

Source: Copyright/used with permission of/courtesy of IEEE.

2.4.2 Measurement Results at X-Band

The designed MS is fabricated using photolithography on top of the RT/duroid 5880 substrate having permittivity 2.2 and thickness 0.762 mm. The fabricated prototype of DASR MS is shown in Figure 2.15(a). The transmission and reflection characteristics of the proposed metamaterial surface are measured in an anechoic chamber using free space measurement technique as reported in [37]. A normal approximated plane wave is incident on the MS using horn antenna WG port and microwave absorbers on the two sides as shown in Figure 2.15(b). The absorption level of

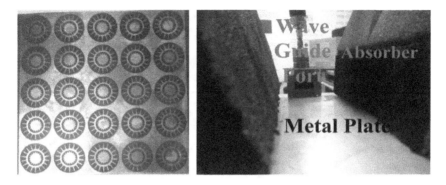

FIGURE 2.15
(a) The fabricated DASR MS and (b) measurement setup for DASR meta-surface.

Source: Copyright/used with permission of/courtesy of IEEE

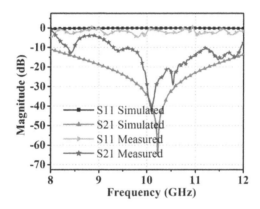

FIGURE 2.16
Measured and simulated transmission and reflection characteristics.

Source: Copyright/used with permission of/courtesy of IEEE.

absorber used is seen to be less than −30 dB. The absorbers are used to avoid interference.

The measured and simulated transmission and reflection characteristics of DASR MS are compared in Figure 2.16. Both are in good agreement except for a very small unwanted change in S_{21} measured plot at about 10.5 GHz and small ripples in S_{11}. This small error could be due to fabrication inaccuracies and misalignment. The proposed MS is observed to be bilateral and polarization-insensitive due to symmetry. This MS is used for antenna radiation characteristic enhancement in the upcoming chapters.

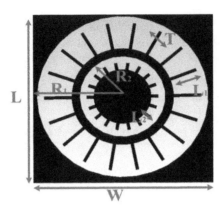

FIGURE 2.17
Rescaled DASR UC with dimension. Dimensions are $L = 11.5$, $W = 11.5$, $R_1 = 5.25$, $R_2 = 1.5$, $L_1 = 1.35$, $L_2 = 0.4$, $T = 0.2$ (all dimensions are in millimetres).

Source: Copyright/used with permission of/courtesy of IEEE.

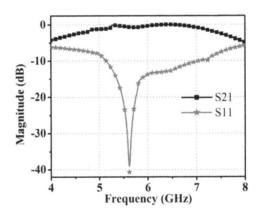

FIGURE 2.18
Simulated transmission and reflection characteristics of new rescaled DASR UC.

Source: Copyright/used with permission of/courtesy of IEEE.

2.4.3 Measurement Results at C-Band

The UC presented in Section 2.3.1 is rescaled for C-band application. The new scaled DASR UC with dimension is shown in Figure 2.17. A 5 × 5 array of new DASR UC is designed and printed on RT/duroid 5880 substrate with permittivity 2.2 and thickness of 0.787 mm (0.0110λ). The size of the UC is only $0.166\lambda \times 0.166\lambda$, where λ is wavelength at lowest pass-band frequency, making the proposed MS ultrathin and compact.

The new DASR UC is simulated using perfect electric and perfect magnetic boundary condition with normal incidence of plane wave on MS. A simulated resonance dip of −40.85 dB is observed at 5.58 GHz as shown in Figure 2.18.

FIGURE 2.19
Free space measurement setup with MS.

Source: Copyright/used with permission of/courtesy of IEEE.

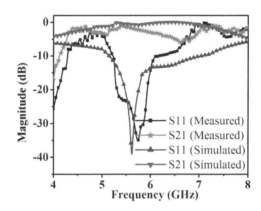

FIGURE 2.20
Measured and simulated transmission and reflection characteristics of MS.

Source: Copyright/used with permission of/courtesy of IEEE.

The material parameters of the proposed new DASR UC are extracted using free space algorithm as discussed in Section 2.3.1. Negative permittivity and permeability characteristics are observed from 4.0 GHz to 6.75 GHz.

The transmission and reflection characteristics of the proposed metamaterial surface are measured in an anechoic chamber using free space measurement technique. The setup used is calibrated first. Direct transmission between two standard C-band WG ports without any air gap is recorded by PNA E8364C VNA. Next, WG ports are placed in free space as shown in the measurement setup (Figure 2.19) and transmission response is recorded by VNA. Now the fabricated new DASR UC MS is placed in far-field region to avoid near field effects, and transmission response is recorded using the setup shown in Figure 2.19. The actual transmission response of MS is the normalized transmission characteristic with respect to free space.

The measured and simulated transmission and reflection characteristics are compared in Figure 2.20. Both are in good agreement except for a very small deviation in measured S_{21} and S_{11}. The measured resonance frequency is shifted to 5.80 GHz. This small error could be due to fabrication inaccuracies and misalignment. A measured transmission bandwidth of 3.75 GHz is observed from 4.25 GHz to 8.0 GHz. This rescaled DASR MS is used in Chapter 5 for designing high-gain MS lens antenna for C-band applications.

2.5 Conclusion

The design of an ultrathin compact miniaturized negative permittivity and permeability MS for C-band, X-band and Ka-band is presented in this chapter. The transmission and reflection characteristics of designed MS are measured using miniaturized free space technique. Measured wide reflection band of about 4 GHz is obtained for X-band applications. The transmission characteristics of C-band MS are measured using free space technique and are found to have transmission bandwidth of 3.75 GHz from 4.25 GHz to 8.0 GHz. Also a compact MS with single-band reflection characteristic and enhanced bandwidth operating in the X-band region is also designed successfully. The designed MSs can be used with any similar frequency band microwave radiator at an optimum height to enhance radiation and transmission characteristics.

In the next chapter, the design of highly miniaturized multiband MPA with significant gain using metamaterials is presented.

References

1. Montrose, M. I., *EMC and the Printed Circuit Board: Design, Theory, and Layout made Simple*, IEEE Press, Institute of Electrical and Electronics Engineers, New York, 1999.
2. Nicolson, A. M. and G. F. Ross, "Measurement of the intrinsic properties of materials by time domain techniques," *IEEE Transactions on Instrumentation and Measurement*, vol. 19, no. 4, pp. 337–382, 1970.
3. Nyshadam, A., C. L. Sibbald, and S. Stuchly, "Permittivity measurements using open-ended sensors and reference liquid calibration-an uncertainty analysis," *IEEE Transactions on Microwave Theory and Techniques*, vol. 40, no. 2, pp. 305–314, 1992.
4. Waser, R., *Nanoelectronics and Information Technology: Advanced Electronic Materials and Novel Devices*, Wiley-VCH, Cambridge, 2003.

5. Weir, W. B., "Automatic measurement of complex dielectric constant and permeability at microwave frequencies," *Proceedings of the IEEE*, vol. 62, no. 1, pp. 33–36, 1974.

6. Ziolkowski, R. W., "Design, fabrication, and testing of double negative metamaterials," *IEEE Transactions on Antennas and Propagation*, vol. 51, no. 7, pp. 1516–1529, 2003.

7. Jerzy Guterman, A., A. Moreira, and C. Peixerio, "Dual-band miniaturized microstrip fractal antenna for a small GSM 1800+ UMTS mobile handset," *Electrotechnical Conference, 2004. MELECON 2004. Proceedings of the 12th IEEE Mediterranean*, vol. 2, IEEE, Dubrovnik, Croatia, pp. 499–501, 2004.

8. Engheta, N. and R. W. Ziolkowski, "A positive future for double negative materials," *IEEE Microwave Theory and Techniques*, vol. 53, no. 4, pp. 1535–1556, April 2005.

9. Baena, J., J. Bonache, F. Martin, R. Sillero, F. Falcone, T. Lopetegi, M. Laso, J. Garcia-Garcia, I. Gil, and M. Portillo, "Equivalent-circuit models for split-ring resonators and complementary split-ring resonators coupled to planar transmission lines," *IEEE Transactions on Microwave Theory and Techniques*, vol. 53, pp. 1451–1461, April 2005.

10. Juan Domingo, B., Jordi Bonache, Ferran Martin, R. Marqués Sillero, Francisco Falcone, Txema Lopetegi, Miguel A. G. Laso et al., "Equivalent-circuit models for split-ring resonators and complementary split-ring resonators coupled to planar transmission lines," *IEEE Transactions on Microwave Theory and Techniques*, vol. 53, no. 4, pp. 1451–1461, 2005.

11. Rajab, Khalid Z., Raj Mittra, and Michael T. Lanagan, "Size reduction of microstrip antennas using metamaterials," *2005 IEEE Antennas and Propagation Society International Symposium*, vol. 2. IEEE, Washington, DC, 2005.

12. Lin, I-H., K. M. K. H. Leong, C. Caloz, and T. Itoh, "Dual-band sub-harmonic quadrature mixer using composite right/left-handed transmission lines," *IEE Proceedings-Microwaves, Antennas and Propagation*, vol. 153, no. 4, pp. 365–375, 2006.

13. Lee, Cheng-Jung, Kevin MKH Leong, and Tatsuo Itoh, "Composite right/left-handed transmission line based compact resonant antennas for RF module integration," *IEEE Transactions on Antennas and Propagation*, vol. 54, no. 8, pp. 2283–2291, 2006.

14. Joseph, C., R. J. Jost. and E. L. Utt, "Multiple angle of incidence measurement technique for the permittivity and permeability of lossy materials at millimeter wavelengths," *IEEE AP-S International Symposium Digest*, pp. 640–643, 1987.

15. Ghodgaonkar, D. K., V. V. Varadan, and V. K. Varadan, "A freespace method for measurement of dielectric constants and loss tangents at microwave frequencies," *IEEE Transactions on Instrumentation and Measurement*, vol.38, pp. 789–793, 1989.

16. Rytting, D., "Advances in microwave error correction techniques," *Hewlett-Packard RF & Microwave Measurement Symposium*, pp. 5954–8378, June 1, 1987.

17. Williams, J., "Accuracy enhancement fundamentals for vector network analyzers," *Microtvave Journal*, vol. 32, pp. 99–114, 1989.

18. Rytting, D., "Let time domain response provide additional insight into network behavior," *Hewlett-Packard RF & Microwave Measurement Symposium*, pp. 5952–6660, March 1985.

19. Barry, W., "A broad-band. automated, strapline technique for simultaneous measurement of complex permittivity and permeability," *IEEE Transactions on Microwave Theory and Techniques,* vol. MTT-34, pp. 80–84, 1986.

20. Chin, G. Y. and E. A. Mechtly, "Properties of materials," in *Reference Datu for Engineering: Radio, Electronics, Computer, Cot17- medications,* ed by E. C. Jorden, Howard W. Sam, Indianapolis, IN.

21. Donecker, B., "Determining the measurement accuracy of the HP8510 microwave network analyzer," *Hewlett-Packard RF & Microwave Measurement Symposium,* Oct. 1984.

22. Nicolson, A. M. and G. F. Ross, "Measurement of intrinsic properties of materials by time domain techniques," *IEEE Transactions on Instrumentation and Measurement,* vol. IM-19, pp. 377–382, Nov. 1970.

23. Ziolkowski, R. W., "Design, fabrication, and testing of double negative metamaterials," *IEEE Transactions on Antennas and Propagation,* vol. 51, no. 7, pp. 1516–1529, July 2003.

24. Weir, W. B., "Automatic measurement of complex dielectric constant and permeability at microwave frequencies," *Proceedings of the IEEE,* vol. 62, pp. 33–36, Jan. 1974.

25. Kadaba, P. K., "Simultaneous measurement of complex permittivity and permeability in the millimeter region by a frequency-domain technique," *IEEE Transactions on Instrumentation and Measurement,* vol. IM-33, no. 4, pp. 336–340, Dec. 1984.

26. Ghodgaonkar, D. K., V. V. Varadan, and V. K. Varadan, "Free-space measurement of complex permittivity and complex permeability of magnetic materials at microwave frequencies," *IEEE Transactions on Instrumentation and Measurement,* vol. 39, pp. 387–394, April 1990.

27. Baker-Jarvis, J., E. J. Vanzura, and W. A. Kissick, "Improved techniques for determining complex permittivity with the transmission/reflection method," *IEEE Transactions on Microwave Theory and Techniques,* vol. 38, pp. 1096–1103, Aug. 1990.

28. Anderson, J. M., C. L. Sibbald, and S. S. Stuchly, "Dielectric measurements using a rational function model," *IEEE Transactions on Microwave Theory and Techniques,* vol. 42, no. 2, pp. 199–204, 1994.

29. Athey, T. W., M. A. Stuchly, and S. S. Stuchly, "Measurement of radio-frequency permittivity of biological tissues with an open-ended coaxial line - Part I," *IEEE Transactions on Microwave Theory and Techniques,* vol. 30, no. 1, pp. 82–86, 1982.

30. Bérubé, D., F. M. Ghannouchi, and P. Savard, "A comparative study of four open-ended coaxial probe models for permittivity measurements of lossy dielectric biological materials at microwave frequencies," *IEEE Transactions on Microwave Theory and Techniques,* vol. 44, no. 10, pp. 1928–1934, 1996.

31. Brady, M. M., S. A. Symons, and S. S. Stuchly, "Dielectric behavior of selected animal tissues *in vitro* at frequencies from 2 to 4 GHz," *IEEE Transactions on Biomedical Engineering,* vol. 28, no. 3, pp. 305–307, 1981.

32. Burdette, E. C., F. L. Cain, and J. Seals, "*In vivo* probe measurement technique for determining dielectric properties at VHF through microwave frequencies," *IEEE Transactions on Microwave Theory and Techniques,* vol. 28, no. 4, pp. 414–427, 1980.

33. Catalá-Civera, J. M., A. J. Canós, F. L. Peñaranda-Foix, and E. de los Reyes Davó, "Accurate determination of the complex permittivity of materials with transmission reflection measurements in partially filled rectangular waveguides," *IEEE Transactions on Microwave Theory and Techniques*, vol. 51, no. 1, pp. 16–24, 2003.
34. Singh, A. K., Mahesh P. Abegaonkar, and Shiban K. Koul, "Wide band low profile ultra-thin meta-surface for X-band application," *2017 IEEE Applied ElectromagneticConference (AEMC)*, Aurangabad, India, pp. 1–2, 2017.
35. Singh, A. K., Mahesh P. Abegaonkar, and Shiban K. Koul, "High-gain and high-aperture-efficiency cavity resonator antenna using metamaterial superstrate," *IEEE Antennas and Wireless Propagation Letters*, vol. 16, pp. 2388–2391, June 2017.
36. Singh, A. K., Mahesh P. Abegaonkar, and Shiban K. Koul, "Ultrathin miniaturized meta-surface for wide band gain enhancement," *2017 IEEE Asia Pacific Microwave Conference (APMC)*, Kuala Lumpar, pp. 383–386, 2017.
37. Engheta, N. and Richard W. Ziolkowski, eds., *Metamaterials: Physics and Engineering Explorations*, John Wiley & Sons, New York, pp. 123–138, 2006.

3

Miniaturization of Microstrip Patch Antennas Using Metamaterials

3.1 Introduction

Miniaturized antennas are the major requirement for microwave mono-lithic integrated circuits to provide proper communication solution. The desire to incorporate multiple frequency bands of operation into personal communication devices has led to much research on reducing the size of antennas while maintaining adequate performance. In addition, antennas and RF front ends, which form the integral part of a wireless system, have not been exempted from the demand for miniaturization. The advancements in the field of material sciences, antenna design techniques, Surface Mount Device lumped components and fabrication technologies have led to the development of many pathways to achieve miniaturization. This in turn has paved the way for researchers in antenna miniaturization to engage actively in adding new dimensions to the field. Due to various technological advance-ments, there are many antenna miniaturization techniques known now, and some of them will be discussed later.

3.2 Antenna Miniaturization Techniques

There are a number of techniques for antenna miniaturization. In the follow-ing section, a brief description of some of these techniques is given.

3.2.1 Antenna Miniaturization Using High Refractive Index Medium

The basic idea behind miniaturization using this technique is that the velocity of wave reduces with an increase in the refractive index of the material. The reduced velocity causes a decrease in the respective propagating wavelength, resulting in a decrease in the electrical size of the antenna. In other words, if

DOI: 10.1201/9781003045885-3

an antenna is designed using a high refractive index medium, then its size will be approximately reduced by a factor of $1/\sqrt{\varepsilon_r \mu_r}$ (where ε_r and μ_r are the relative permittivity and permeability of the medium, respectively). Similarly, a high dielectric constant, high relative permeability or both may be used to achieve miniaturization. It is also observed that a good impedance match is possible if ε_r and μ_r (the relative permittivity and permeability of the medium) are equal [1]. Increasing refractive index may increase the loss tangent of the medium, resulting in a decrease in the broadside radiated gain [2–3]. Much attention has been directed towards high dielectric constant materials such as ceramics owing to their low loss characteristics, in spite of the fact that antennas on such substrates exhibit low impedance bandwidth, low radiation efficiency and difficulty in impedance matching. Loading the substrate using superstrates has been effectively used in planar structures such as MPAs, printed dipoles and spiral antennas [4–5].

3.2.2 Antenna Miniaturization by Shaping

Miniaturization by shaping is one of the most extensively used techniques for designing small antennas. The idea behind shaping is to increase the surface current propagation path length on the surface of the antenna, resulting in increase in resonant length of the antenna. This results in miniaturization of the patch antenna. For example, in a rectangular patch antenna, the current flows from one edge to the other. Addition of slots, however, changes the direction which requires the current to take a longer path. Thus, the resonant frequency of the patch is reduced.

Shaping includes various geometrical modifications of the antenna such as bending, meandering, folding, slot loading, fractal folding, etc. Bending and folding are usually applied to antennas which are 3D and shaped similar to a monopole. The printed inverted-F (PIFA) is one such antenna that is widely used in handheld wireless devices [6]. Folding is also widely used for miniaturization of ultra-wideband (UWB) monopole antennas [7]. These techniques reduce the size of the antenna in the desired direction at the expense of an increase in size in another direction or increase in volume of the antenna [7]. Adding slots is another popular technique used widely in planar antennas such as MPAs. The addition of slots in most cases increases the resonant length of the antenna, thereby reducing the resonant frequency of operation [8].

3.2.3 Antenna Miniaturization by Lumped Element Loading

In antenna applications where shaping may not be considerable because of some other device aspects, the effects of addition of slots and meandering may be achieved by including equivalent lumped elements to the antenna circuit, generally capacitors and inductors. The addition of proper value of lumped component in series or in parallel results in significant miniaturization of the antenna. However, in this technique of miniaturization, sometimes gain may decrease. In [9], a coplanar waveguide (CPW), folded slot antenna with capacitive loading using chip capacitors has been reported. A 22 per cent size reduction is achieved using this technique at the expense of 2.7-dB reduction in the measured gain and 11 per cent in 10-dB return loss bandwidth of the antenna. Another similar example for capacitive loading has been reported in [10]. Here, the authors have studied the effect of miniaturization of loop antenna with multiple periodically arranged shunt capacitances. The results indicate seven-fold size reduction of the original antenna, with a 7-dB reduction in the gain. In [11], lumped inductors for miniaturizing loop antenna have also been demonstrated.

3.2.4 Antenna Miniaturization Using Metamaterial Loading

In recent years, artificially engineered materials such as metamaterials have received lot of attention in the field of antenna miniaturization. These materials, made up of periodic sub-wavelength elements, generate unique material properties utilizing element resonances. Examples of such materials are electromagnetic band-gap (EBG) materials [12], DNG materials, artificial magnetic conductors (AMCs), frequency-selective surfaces (FSSs) and metamaterials. The application of metamaterial for miniaturization is found to be of great importance. The idea behind using metamaterial for doing miniaturization is by loading the radiating patch with capacitive and inductive loads in series or shunt. These inductive and capacitive loads can be modelled by designing appropriate metamaterial UCs. It has been shown in [13] that EBG can be used to reduce the distance between the radiating element and its ground effectively. These materials also find applications in surface wave reduction of patch antennas [14]. As shown in [15], composite right-/left-handed (CRLH) transmission line antennas are other candidates for antenna miniaturization in both leaky wave and resonant antenna structures. However, a loss of energy is one of the fundamental limitations associated with metamaterials. It has been shown in [16] that any effort to reduce the loss in metamaterial reduces the artificial property (negative refractive index) associated with it. This follows from the fundamental principles of causality obtained from the dispersion relations of the material.

This chapter presents the design and development of highly miniaturized double-band and triple-band MPAs. The principle of the proposed patch antenna element is based on adding series or shunt impedance to decrease the half-wavelength resonance frequency, thus reducing the electrical size of the proposed patch antenna. First, a miniaturized double-band patch antenna with significant broadside and backside gain is designed. Next, the same approach is extended to design a triple-band highly miniaturized significant broadside and backside gain. The patch antenna is fabricated and experimentally characterized by measuring reflection coefficient and radiation characteristic. The simulated and measured results are in good agreement.

3.3 Highly Miniaturized Dual-Band Patch Antenna Loaded with Metamaterial Unit Cell

In this section, a highly miniaturized low-profile dual-band patch antenna loaded with metamaterial UC is discussed. The metamaterial UC consists of an interdigital capacitor (IDC) and a novel modified circular slot ring resonator (MCsRR). The novel MCsRR is a low-frequency compact circular slot ring resonator (CsRR). The proposed antenna has dual-band characteristics with reasonable gain of 2.84 dBi at first centre frequency of 3.15 GHz with bandwidth of 150 MHz and 3.86 dBi at second centre frequency of 5.28 GHz [17].

3.3.1 Antenna Design and Working Principle

Initially, a square MPA of dimension 19 mm × 19 mm is designed on Neltec substrate with permittivity (ε_r) = 2.2 and thickness 0.762 mm with simple ground as shown in Figure 3.1(a) with return loss S_{11} in curve "*a*" of Figure 3.2. The antenna is resonating at 10.3 GHz. The electrical size of a square patch antenna is miniaturized using novel metamaterial UC. A metamaterial UC consists of IDC loaded with a novel MCsRR. The square patch antenna is loaded with IDC of finger length "L_c" of 8.75 mm with finger thickness "G_c" of 1 mm and with air gap "P_c" of 0.2 mm with simple ground. IDC loading causes change in resonance frequency of previous antenna to 7.50 GHz as in curve "*b*" of Figure 3.2. Resonating frequency is further reduced to 3.20 GHz by slotting the ground plane with a circular slot of radius "R_1" of 6 mm, as in Figure 3.1(c), and patch as in Figure 3.1(b) with S_{11} in curve "*c*" of Figure 3.2. Next, the novel MCsRR is added to the ground of the antenna to get the second desired resonating band at 5.2 GHz, as in Figure 3.1(d) and in curve "*d*" of Figure 3.2.

Novel MCsRRs consist of a metallic circle with radius "R" = 1.50 mm with extended metallic finger having length "r" on it. The thickness of the metallic finger on MCsRR is "t" = 0.30 mm, as given in Figure 3.3. The circular metallic section with extended metallic finger adds a finite inductance "L_m" and capacitance "C_s and C_m" (Figure 3.6). This type of metallic structure increases the current path and leads to multi-mode propagation in the antenna. If propagating modes are far enough in frequency, then multiband operation

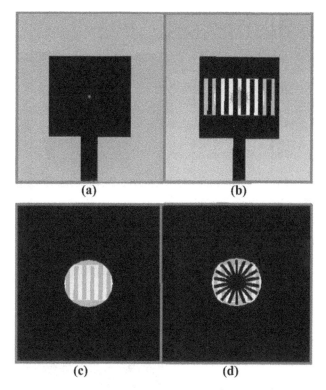

FIGURE 3.1
Design steps of proposed antenna loaded with metamaterial UC. (a) Square patch antenna with simple ground, (b) IDC-loaded square patch antenna with simple ground, (c) IDC-loaded square patch antenna with circular slot in ground and (d) IDC-loaded square patch antenna with an MCsRR in ground.

Source: Copyright/used with permission of/courtesy of Wiley.

occurs. The detailed dimensions of the proposed structure are shown in Figure 3.3.

IDC loaded with MCsRR behaves as a metamaterial UC. The model of metamaterial UC is simulated and model is fabricated to study the metamaterial pass and stop-band behaviour as reported in [18]. The fabricated model is shown in Figure 3.4. Simulated and measured scattering parameter results are in good agreement, as shown in Figure 3.5. Zero reflection at some frequency indicates zero shunt admittance of metamaterial UC. The resonant tank circuit of MCsRR is open-circuited; hence, impedance matching can be done easily. Zero transmission at some frequency indicates infinite shunt admittance of metamaterial UC. The resonant tank circuit of MCsRR is short-circuited; hence, impedance matching cannot be done. IDC loaded with MCsRR behaves as a band-pass filter for both radiating bands of proposed antenna as shown in Figure 3.5.

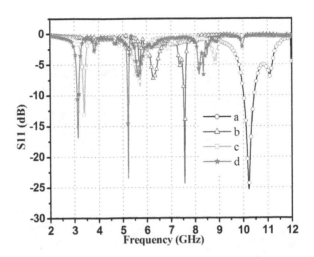

FIGURE 3.2
Return loss (S_{11}) of designing steps a, b, c and d of proposed antenna.

Source: Copyright/used with permission of/courtesy of Wiley.

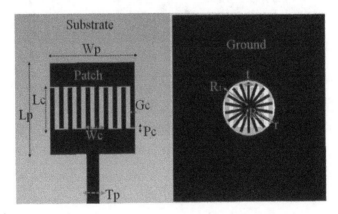

FIGURE 3.3
Proposed metamaterial UC-loaded microstrip patch antenna; (a) top view and (b) back view. Dimensions are $L_p = 19.00$, $W_p = 19.00$, $L_c = 8.75$, $W_c = 3.00$, $G_c = 1.00$, $P_c = 0.20$, $T_p = 2.76$, $R_1 = 6.00$, $r = 4.50$ and $R = 1.50$ (all dimensions are in millimetres).

Source: Copyright/used with permission of/courtesy of Wiley.

The equivalent circuit model of patch antenna loaded with metamaterial UC shown in Figure 3.6 consists of series equivalent inductance "L_t" due to pre-section of transmission line, series capacitance "C_i" due to IDC, shunt LC network "L_m, C_s and C_m" due to the MCsRR being slotted in the ground plane. Equivalent capacitance "C_s" is due to split air gap between MCsRR metallic

FIGURE 3.4
Fabricated model of metamaterial UC to measure transmission and reflection characteristics;
(a) top view and (b) back view.

Source: Copyright/used with permission of/courtesy of Wiley.

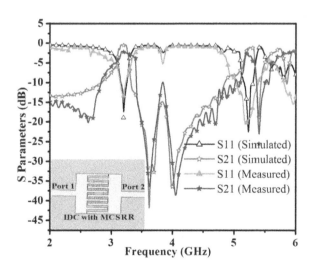

FIGURE 3.5
Simulated and measured S-parameters of transmission model of metamaterial UC.

Source: Copyright/used with permission of/courtesy of Wiley.

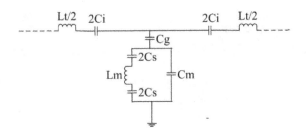

FIGURE 3.6
Equivalent circuit model of MCsRR loaded proposed patch antenna.

Source: Copyright/used with permission of/courtesy of Wiley.

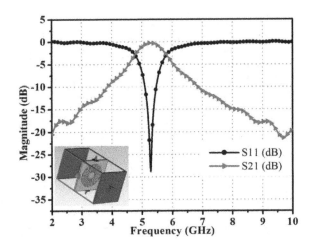

FIGURE 3.7
Simulated return loss and transmission parameters of MCsRR.

Source: Copyright/used with permission of/courtesy of Wiley.

finger and ground plane. Equivalent inductance "L_m" is due to metallic finger on MCsRR, and equivalent capacitance "C_m" is due to capacitance between metallic fingers. Both series and shunt networks are separated by capacitance "C_g" due to the ground plane.

The MCsRR UC is simulated under PEC and PMC boundary in free space with normal incidence of wave on the sample using WG port. Reflection and transmission parameters of sample are plotted in Figure 3.7. It is observed that MCsRR resonates at 5.21 GHz.

The surface current on patch antenna and MCsRR at 3.15 GHz and 5.28 GHz are shown in Figure 3.8(a), (b), (c) and (d), respectively. It is seen that surface current distribution is concentrated on IDC structure on the top of patch and on MCsRR fingers on the ground plane. The surface current distribution mentioned earlier indicates

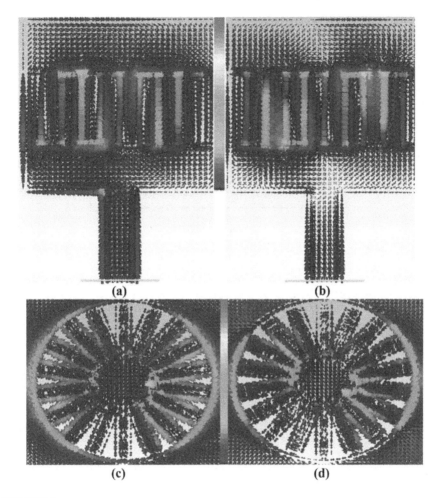

FIGURE 3.8
Simulated current distribution on proposed patch antenna (a) Patch at 3.15 GHz (b) Patch at 5.28 GHz (c) MCsRR at 3.15 GHz and (d) MCsRR at 5.28 GHz.

Source: Copyright/used with permission of/courtesy of Wiley.

that source of first and second resonant bands is IDC and MCsRR, respectively.

Variations in radius "R" of proposed MCsRR cause significant change in the second resonant band. As "R" increases from 1 mm to 3 mm, resonant frequency of the second band moves from 5.2 GHz towards the higher frequency side, as shown in Figure 3.9. Variation in "R" causes change in length of metallic strip "r". Due to this, overall equivalent inductance "L_m" and capacitance "C_s and C_m" added by the MCsRR to the circuit also change, causing a change in the second resonant band. Figure 3.7 also predicts that the second band is generated by the MCsRR. Hence, the second resonant band can be

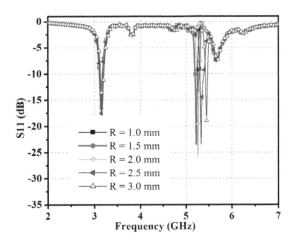

FIGURE 3.9
Reflection coefficient due to variations in the outer radius "*R*" of the MCsRR.

Source: Copyright/used with permission of/courtesy of Wiley.

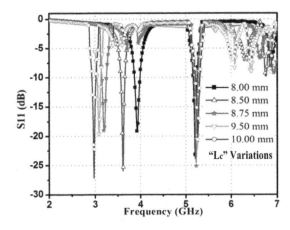

FIGURE 3.10
Reflection coefficient due to variations in finger length "L_c" of interdigital capacitor from 8.00 mm to 10.00 mm.

Source: Copyright/used with permission of/courtesy of Wiley.

controlled by "*R*". The second resonant band is induced by slotting MCsRR in the ground as shown in Figure 3.1(d) and curve "*d*" of Figure 3.3.

Variations in finger length "L_c" of IDC cause significant change in the first resonant band. Change in the second resonant band due to variations in "L_c" is insignificant. When "L_c" increases from 8 mm to 10 mm, the first resonant band moves toward the lower frequency side, as shown in Figure 3.10. Due to an increase in finger length of IDC, overall series capacitance "C_i" changes and it causes change in the first resonant band. Hence, the first resonant band can be controlled by "L_c". A variation in "t", thickness of metallic strip, does not contribute to significant change in either of the resonant bands.

3.3.2 Experimental Results

In this section, the fabricated antenna's measured results are discussed. The prototype antenna is fabricated, and measurements are carried out using VNA in an anechoic chamber. IDC with finger length "L_c" of 8.75 mm and finger thickness "G_c" of 1 mm with air gap "P_c" of 0.2 mm is printed on a square patch antenna of size 19 mm × 19 mm. IDC finger length is optimized to 8.75 mm to obtain a resonance band at 3.15 GHz and is printed on the patch antenna. A novel MCsRR is slotted on the ground plane of the antenna to generate multiple modes for dual-band operation. The fabricated prototype antenna is shown in Figure 3.11. The measured and simulated return loss (S_{11}) of prototype antenna loaded with metamaterial UC is presented in Figure 3.12. As observed, the measured results are in good agreement with simulated results.

The proposed patch antenna has electrical size of 0.199λ × 0.199λ (19 mm × 19 mm), where λ is associated with first resonance frequency of 3.15 GHz. Conventional patch antenna size is miniaturized by 60.10

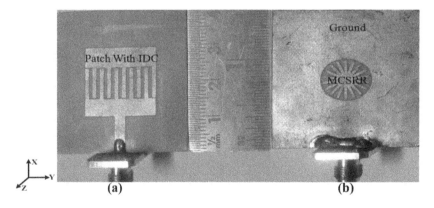

FIGURE 3.11
Fabricated metamaterial UC-loaded microstrip patch antenna; (a) top view and (b) back view.

Source: Copyright/used with permission of/courtesy of Wiley.

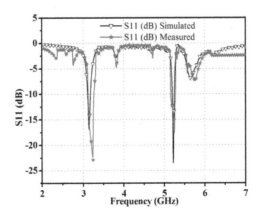

FIGURE 3.12
Simulated and measured return loss characteristics of prototype antenna.

Source: Copyright/used with permission of/courtesy of Wiley.

per cent. The IDC structure with increased interdigital finger length causes electrical size reduction of antenna as compared to the conventional patch antenna. The measured 10-dB return loss bandwidth is 150 MHz for first resonant band extending from 3.12 GHz to 3.27 GHz; the second resonant band extends from 5.16 GHz to 5.30 GHz with bandwidth of 140 MHz.

The radiation pattern of fabricated prototype antenna is measured in an anechoic chamber. The E-plane (XZ) and H-plane (YZ) measured radiation patterns for both resonating bands are plotted in Figure 3.13. The proposed antenna supports back-radiation due to MCsRR. The antenna has near-omnidirectional pattern in H-plane and has near-dipolar radiation pattern in E-plane for both resonating bands. It is observed that the back-radiation is independent of size of the ground plane.

The measured gain of the prototype patch antenna at different frequencies is plotted in Figure 3.14. The gain plot indicates gain of 2.84 dBi at the first centre frequency of 3.15 GHz and 3.86 dBi at the second centre frequency of 5.28 GHz. Although electrical size as well as physical size of the patch antenna is very small as compared to the conventional patch antenna, the proposed antenna shows significant high gain at both the frequencies. Several reported dual-band antennas are compared with the proposed antenna in Table 3.1. The proposed antenna occupies the smallest area with significant gain as compared to other reported dual-band antennas.

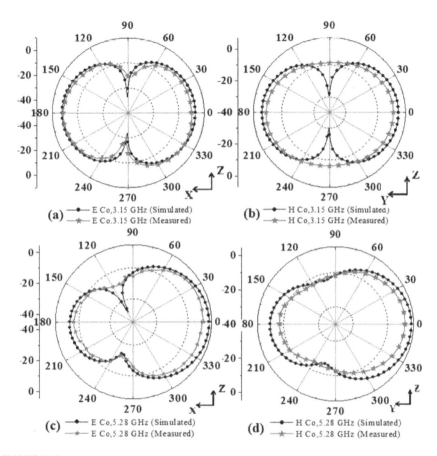

FIGURE 3.13

Radiation pattern of proposed antenna (a) E-plane at 3.15 GHz, (b) H-plane at 3.15 GHz, (c) E-plane at 5.28 GHz and (d) H-plane at 5.28 GHz.

Source: Copyright/used with permission of/courtesy of Wiley.

A highly miniaturized low-profile dual-band monopole MPA loaded with a novel metamaterial UC is successfully proposed, simulated, fabricated and tested for multiple applications. The proposed antenna is miniaturized by 60.10 per cent with significant calculated antenna gain. MCsRR etched in the ground plane causes different current distribution and hence multiple band operation with 10-dB return loss bandwidth of 150 MHz and 140 MHz with moderate gain of 2.84 dBi for the first centre frequency and 3.86 dBi for the second centre frequency. In the next section, miniaturized triple-band patch antenna design is presented.

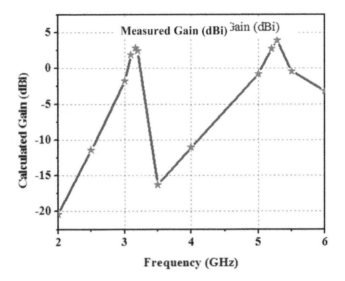

Measured Gain (dBi)

FIGURE 3.14
Measured gain versus frequency of the prototype antenna.

Source: Copyright/used with permission of/courtesy of Wiley.

TABLE 3.1

Comparison of Proposed Antenna with Several Dual-Band Reported Antennas

Ref.	Size (mm)	Frequency Band (GHz)	Operating Bandwidth (MHz)	Miniaturization (%)
[19]	22.56 × 36.40	1.79 & 2.038	380 & 330	38.00
[20]	32.00 ×15.00	1.34 & 2.87	310 & 270	37.00
[21]	55.00 × 55.00	2.45 & 5.20	350 & 300	10.10
[22]	20.00 × 19.00	2.0 & 3.40	180 & 200	48.40
Proposed	19.00 × 19.00	3.15 & 5.28	150 & 140	60.10

3.4 Triple-Band Miniaturized Patch Antenna Loaded with Metamaterial Unit Cell

In this section, an optimal triple-band miniaturized MPA loaded with planar metamaterial UC is proposed. The fundamental principle for achieving miniaturization is the same as discussed in Section 3.2. The proposed prototype patch antenna operates in the frequency bands of 5.28 GHz, 5.78 GHz and 6.98 GHz, having significant measured antenna gain [23].

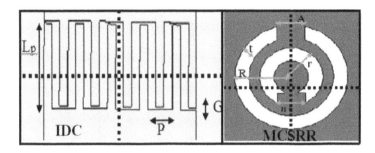

FIGURE 3.15
Metamaterial UC (a) interdigital capacitor and (b) modified circular split-ring resonator. The dimensions are L_p = 7.10, P = 3.00, G = 1.00, A = 4.00, a = 4.00, b = 2.37, R = 7.75, r = 4.00 and t = 2.00. (all dimensions are in millimetres).

Source: Copyright/used with permission of/courtesy of IEEE.

3.4.1 Antenna Design and Working Principle

A square patch antenna of dimension 19 mm × 19 mm on Neltec substrate having permittivity (ε_r) 2.2, substrate thickness 0.762 mm and metal thickness 0.017 mm is loaded with IDC on the patch, so that the overall electrical length of the patch can be reduced. The IDC adds some finite capacitance in series with patch antenna equivalent circuit. Due to this, the overall capacitance of the antenna can be controlled and the effective electrical length of patch antenna can be reduced. The capacitance of IDC can be controlled by finger length of the IDC.

The IDC is designed to reduce the resonant frequency from 10.5 GHz to 7 GHz. The finger length, air gap and periodicity of the finger on IDC are parameterized and optimized to get proper value of the capacitance. The empirical formula for calculating capacitance of IDC is used here. The dimension of IDC is optimized to get the desired frequency response. Finger length (L_p) of 7.10 mm, air gap (G) of 1 mm and periodicity (P) of 3 mm are considered for the designed IDC as shown in Figure 3.15(a). The metamaterial UC consists of IDC loaded with novel modified circular split ring resonator (MCsRR). MCsRR is a modified version of circular split ring resonator (CsRR) as shown in Figure 3.15(b). The split gap "*a*" of simple CSRR is modified to get different value of capacitance and inductance. The maximum surface current concentration is observed near inner and outer split gaps. The outer split gap "*A*" of MCsRR is same as that of CSRR, but inner split gap is modified to enhance current concentration and current path. This modification leads to change in inductance and capacitance in the equivalent circuit. The effect of penetration angle on inner split of MCsRR is also studied and explained.

The equivalent circuit model of patch antenna loaded with metamaterial UC, given in Figure 3.16, consist of series *LC* combination, where equivalent inductance (L_t) is due to pre-section transmission line and equivalent capacitance (C_i) is due to IDC. An extra shunt equivalent *LC* resonant tank circuit is due to slotted novel (MCsRR). Finger length of IDC and dimensions of slotted novel MCsRR can control resonating modes of the patch antenna loaded with metamaterial UC.

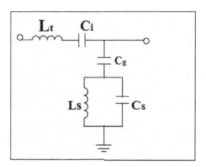

FIGURE 3.16
Equivalent circuit model of proposed prototype antenna.

Source: Copyright/used with permission of/courtesy of IEEE

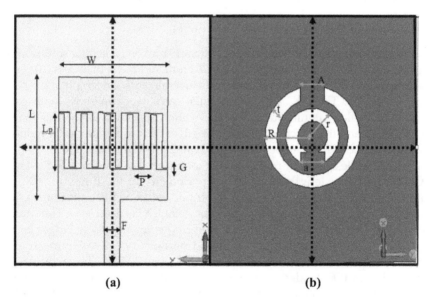

$\qquad$ (a) $\qquad$ (b)

FIGURE 3.17
Proposed prototype patch antenna; (a) top view and (b) back view. Dimensions are $L = 19.00$, $W = 19.00$, $L_p = 7.10$ and $F = 2.76$. (all dimensions are in millimetres).

Source: Copyright/used with permission of/courtesy of IEEE.

The designed patch antenna with slotted IDC on the patch is loaded with a planar metamaterial UC to obtain multiple bands of antenna as well as miniaturization of the patch antenna. The proposed prototype antenna is shown in Figure 3.17. The UC consists of IDC and novel modified circular SRR (MCsRR). This MCsRR is slotted in the ground plane of patch antenna as in Figure 3.17(b). In the designed antenna system, IDC adds series capacitance, and MCsRR generates shunt inductance and capacitance. The equivalent circuit of this type of patch antenna is reported in [24] and is shown in Figure 3.16. The equivalent

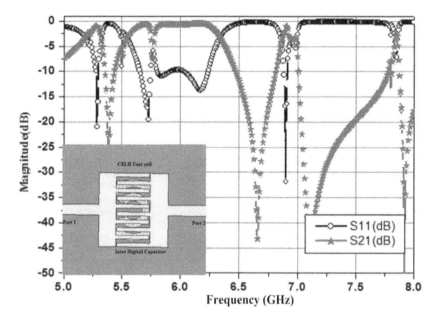

FIGURE 3.18
Transmission and reflection characteristics of transmission model of UC.

Source: Copyright/used with permission of/courtesy of IEEE.

circuit of this model consists of one series *LC* tank circuit and another shunt *LC* tank circuit. Both of these circuits are connected with a capacitor, which is the equivalent capacitance due to ground and metallic patch of antenna "C_g". The shunt *LC* tank circuit is due to novel MCsRR slot in the ground plane. Depending on the values of capacitor and inductor, single to multiple modes can be excited in the proposed prototype patch antenna.

The metamaterial UC is simulated to study the characteristics and metamaterial properties. The transmission and reflection characteristics of transmission model of metamaterial UC are shown in Figure 3.18, gives some near-zero transmission as well as near-zero reflection. Zero reflection at some frequency indicates zero shunt admittance of the UC. Because the resonant tank circuit of MCsRR is open-circuited, impedance matching can be done. Zero transmission at some frequency indicates infinite shunt admittance of the UC. Because the resonant tank circuit of MCsRR is short-circuited, impedance matching cannot be done.

3.4.2 Antenna Design Analysis

Surface current distribution on the IDC and MCsRR is studied at resonant frequencies of 5.28 GHz, 5.78 GHz and 6.98 GHz, and the results are plotted in Figure 3.19. As the patch antenna is loaded with IDC and also with an

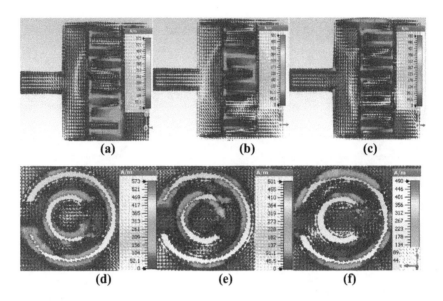

FIGURE 3.19
Simulated surface current distribution.

Source: Copyright/used with permission of/courtesy of IEEE.

MCsRR slot in the ground plane, three bands are generated due to mutual coupling and change in current density.

The surface current density at the first resonant frequency is mainly on the right side of IDC as shown in Figure 3.19(a), on left side of IDC as shown in Figure 3.19(b) at second resonant frequency and on left most side of patch arm and IDC at third resonant frequency as shown in Figure 3.19(c). Different surface current distributions are observed on MCsRR at all three resonant frequencies.

The dimensions of inner split, outer split and IDC finger length are changed and effect on reflection coefficient is studied. It is observed that changing "*a*" width of inner split changes the slot penetration angle of the inner split from 0 to 90 degrees. The current concentration changes drastically. Due to this, the first and third resonance bands remain unaffected, but second resonance band of prototype antenna changes as shown in Figure 3.20. As "*a*" increases, the second resonating band moves towards the higher frequency side and resonance strength decreases, and at about 6 GHz very weak resonance is observed. The second resonating band can be controlled by changing slot penetration angle as well as slot width of the inner smaller ring.

It is observed that by changing "*A*" (slot length of outer rings), slot thickness is unaffected and slot width changes. Due to this, the second and third resonance bands are affected slightly (almost unchanged), but

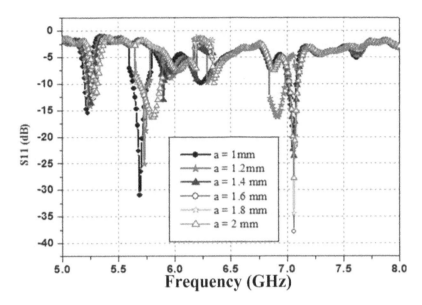

FIGURE 3.20
Reflection coefficient with variations in "*a*" inner slot penetration of MCsRR.

Source: Copyright/used with permission of/courtesy of IEEE.

the first band changes very fast. The first resonating band can be controlled by slot length of the outer ring. By changing the finger length of IDC (L_p), change in series capacitance will occur. Due to this, the third band of antenna changes, whereas the other two bands are unchanged as shown in Figure 3.21. So the third band can be controlled by finger length of the IDC.

3.4.3 Experimental Results

The prototype of proposed patch antenna is fabricated on Neltec substrate ($\varepsilon_r = 2.2$) with a thickness of 0.762 mm. Figure 3.22 shows the proposed fabricated prototype antenna. The IDC finger length is optimized to 7.10 mm to obtain resonance band at 6.98 GHz, and the structure is printed on to the patch antenna. The novel MCsRR is slotted on the ground plane of the antenna as shown in Figure 3.22(b). The measured and simulated S_{11} (return loss) of the antenna is plotted in Figure 3.23. As observed, the measured characteristics are in good agreement with the simulated results within experimental error. The proposed patch antenna has electrical size of $0.334\lambda \times 0.334\lambda$ (19 mm × 19 mm), where λ is associated with first resonance frequency of 5.28 GHz. Conventional patch antenna size is miniaturized by 49.42 per cent.

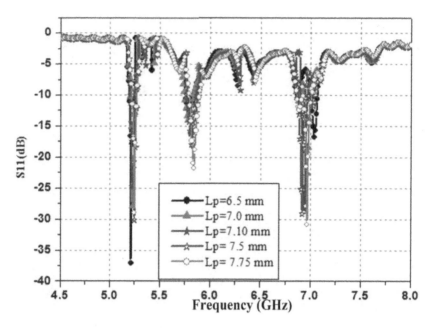

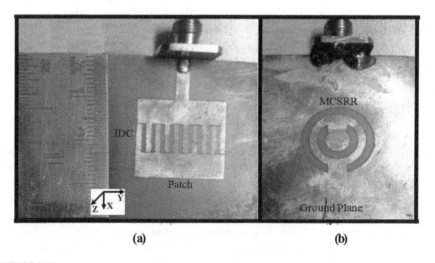

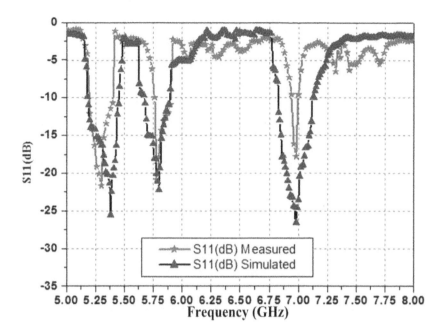

FIGURE 3.23
Simulated and measured reflection coefficient (S_{11}) of the prototype antenna.

Source: Copyright/used with permission of/courtesy of IEEE.

The metamaterial structure with increased interdigital finger length causes electrical size reduction of patch antenna as compared to the conventional patch antenna.

The radiation patterns of fabricated prototype antenna are measured in an anechoic chamber. The radiation pattern of the antenna in all the three bands is plotted in Figure 3.24. The E- and H-planes (XZ and YZ planes) radiation patterns indicating vertical and horizontal polarization are plotted in Figure 3.24. As observed, the proposed antenna has near-omnidirectional radiation pattern for horizontal polarization for all the three bands.

The measured gain of the prototype patch antenna at different frequencies is plotted in Figure 3.25. The gain plot indicates gain of 1.5 dBi at first centre frequency of 5.28 GHz, 3.5 dBi at second centre frequency of 5.78 GHz and 3.80 dBi at third centre frequency of 6.98 GHz. Although electrical size as well as physical size of the patch antenna is very small as compared to a conventional patch antenna, gain of the proposed patch antenna at these frequencies is significantly large. The proposed antenna

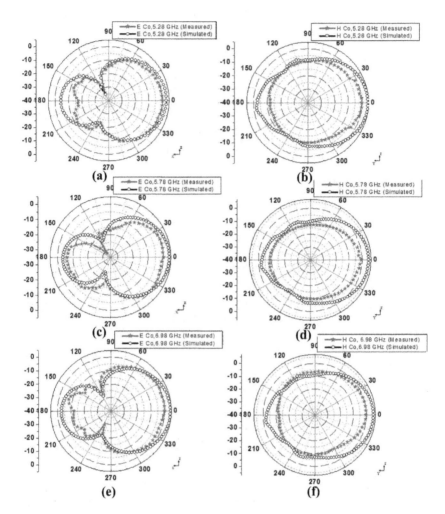

FIGURE 3.24
Radiation pattern of the proposed antenna.

Source: Copyright/used with permission of/courtesy of IEEE.

has three operating bands with first band from 5.19 GHz to 5.41 GHz, second band from 5.76 GHz to 5.92 GHz and third band from 6.94 GHz to 7.00 GHz.

An optimal triple-band MPA loaded with planar metamaterial UC is successfully demonstrated. The proposed antenna achieves miniaturization of about 49.42 per cent as compared to the conventional patch antenna. Although size of the antenna is quite small, gain of antenna in all three bands is quite large.

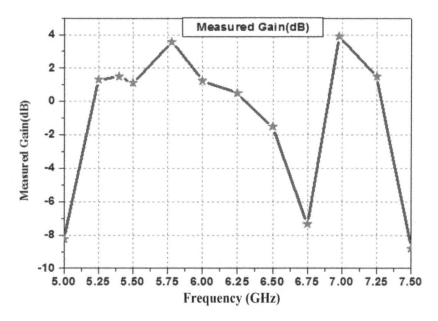

FIGURE 3.25
Measured gain versus frequency of the proposed antenna.

Source: Copyright/used with permission of/courtesy of IEEE.

3.5 Miniaturized Multiband Microstrip Patch Antenna Using Metamaterial Loading

In this section, a highly miniaturized significant gain multiband patch antenna loaded with a modified double circular slot ring resonator (MDCsRR) metamaterial UC is presented. The miniaturization is achieved using inductive and capacitive loading of the radiating patch as reported in [25]. The electrical size of the proposed antenna is miniaturized by about 68.83 per cent as compared to the conventional patch antenna operating at first resonating frequency. The miniaturized patch antenna has significant measured antenna gain [25].

3.5.1 Antenna Design and Working Principle

In this section, the antenna design steps are explained. All the simulations are done using CST microwave studio software. Initially, a MPA of dimensions 19 mm × 19 mm is designed on Neltec substrate having permittivity $(\varepsilon_r) = 2.2$ and thickness 0.762 mm, as shown in Figure 3.26(a), with a simple metallic ground. The antenna resonates at 10.15 GHz (curve "a" of Figure 3.27). To reduce resonant frequency "f_o", patch antenna is further loaded by IDC with simple ground [Figure 3.26(b)]. The resonant frequency is reduced

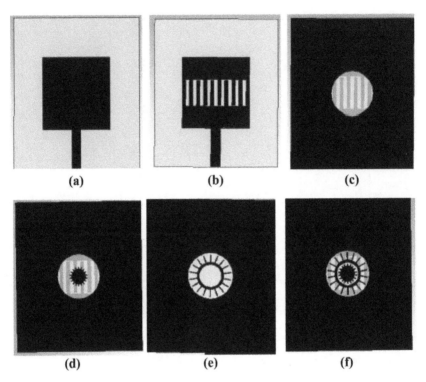

FIGURE 3.26
Design steps of proposed antenna loaded with metamaterial UC. (a) Square patch antenna with simple ground plane, (b) IDC-loaded square patch antenna with simple ground plane, (c) IDC-loaded square patch antenna with circular slot in the ground plane, (d) IDC-loaded square patch antenna with inner slot ring in the ground plane, (e) IDC-loaded patch antenna with outer slot ring only in the ground plane and (f) IDC-loaded patch antenna with MDCsRR in the ground plane.

Source: Copyright/used with permission of/courtesy of EMW.

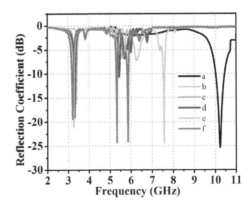

FIGURE 3.27
Reflection coefficient (S_{11}) for different proposed antenna structures a, b, c, d, e and f shown in Figure 3.26.

Source: Copyright/used with permission of/courtesy of EMW.

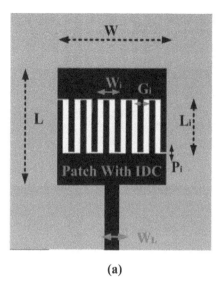

(a)   (b)

FIGURE 3.28

Proposed metamaterial UC-loaded MPA; (a) top view and (b) back view. Dimensions are $L = 19.00$, $W = 19.00$, $L_i = 8.60$, $W_i = 3.00$, $G_i = 1.00$, $P_i = 0.25$, $W_L = 2.76$, $R = 3.52$, $r = 1.98$, $R_1 = 6.00$, $L_1 = 0.50$, $L_2 = 1.94$, $T = 0.52$ and $t = 0.20$ (all dimensions are in millimetres).

Source: Copyright/used with permission of/courtesy of EMW.

to 7.57 GHz (curve "*b*" of Figure 3.27). To reduce resonant frequency "f_o" further, patch antenna with IDC is loaded with a circular defect structure of radius $R_1 = 6.00$ mm in the ground plane [Figure 3.26(c)]. This causes a major shift in the resonant frequency to 3.39 GHz (curve "*c*" of Figure 3.27). To generate the second resonating band, the structure of Figure 3.26(c) is modified to include inner CsRR [Figure 3.26(d)]. We observe the first band at 3.31 GHz and second band at 5.45 GHz as shown in Figure 3.27 (curve "*d*"). When the combination of patch antenna [Figure 3.26(b)] and ground plane structure [Figure 3.26(c)] is further loaded with outer CsRR only [Figure 3.26(e)], we observe the first band at 3.31 GHz and the second band at 5.85 GHz [Figure 3.27 curve (e)]. The aforementioned geometry loaded with MDCsRR, as shown in Figure 3.26(f), generates triple-band operation. Square patch antenna loaded with IDC and MDCsRR generates three resonant bands with resonating frequency at 3.20 GHz, 5.40 GHz and 5.80 GHz, respectively, as depicted in Figure 3.27 (curve "*f*"). In Figure 3.26, the black colour represents the metallic section of the proposed design. The loading of patch antenna with IDC and MDCsRR causes a major shift in the resonant frequency from 10.15 GHz to 3.20 GHz, resulting in miniaturization of 68.83 per cent as compared to a conventional patch antenna size (31.25 mm × 37.05 mm × 0.762 mm) operating at 3.20 GHz.

The change in resonant frequency of IDC-loaded patch antenna with ground slotting depends on mutual coupling between IDC and the slotted MDCsRR geometry. The prototype MPA is loaded with MDCsRR.

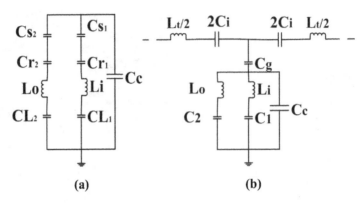

FIGURE 3.29
Equivalent circuit of (a) MDCsRR and (b) patch antenna loaded with IDC and MDCsRR.

Source: Copyright/used with permission of/courtesy of EMW.

MDCsRR metamaterial UC is shown in Figure 3.28 along with dimensions of the antenna structure. IDC is a capacitive load added in series with patch antenna equivalent circuit, where capacitance due to IDC "C_i" depends on finger length "L_i", finger width "G_i" and air gap "P_i". MDCsRR is a compact, low-frequency CsRR. The detailed circuit model of MDCsRR is given in Figure 3.29(a), which consists of several capacitances and inductances. Inner ring resonators have equivalent inductance "L_i" due to inner metallic slot and inner ring, "C_{s_1}" capacitance is due to inner metallic slot of length "L_1" and outer ring conductor, "C_{r_1}" capacitance is due to inner ring and outer ring conductor only, "C_{L_1}" is mutual capacitance due to inner metallic fingers only. Outer ring resonators have equivalent inductance "L_o" due to outer metallic slot and outer ring, "C_{s_2}" capacitance is due to outer metallic slot of length "L_2" and ground plane, "C_{r_2}" capacitance is due to outer ring and ground plane only, "C_{L_2}" mutual capacitance is due to outer metallic fingers only. The equivalent circuit model of patch antenna loaded with IDC and MDCsRR consists of LC series tank circuit and shunt LC tank circuit separated by capacitance due to ground "C_g" as shown in Figure 3.29(b). The shunt LC tank circuit consists of one series combination of "L_o" and "C_2", where "C_2" is total equivalent capacitance due to outer slot ring resonator. Another shunt LC tank circuit consists of series combination of "L_i" and "C_1", where "C_1" is total equivalent capacitance due to inner slot ring resonator. Series LC tank circuit consists of series combination of "L_t" and "C_i", where "L_t" is inductance per unit length of the transmission line. The shunt capacitance "C_c" is capacitance due to a metallic disc of radius "$r - \dfrac{c}{2}$" surrounded by a ground plane at a distance of "c" from its edge as given in [24]. The equivalent circuit in Figure 3.29(b) is obtained by combining equivalent circuit due to IDC, MDCsRR and patch antenna. The circuit parameters can be calculated using conventional analytical approach, as given in [26]. The full-wave EM simulation of the complete structure using CST microwave studio and circuit

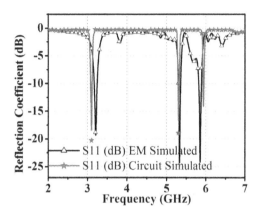

FIGURE 3.30
Comparison of circuit- and EM-simulated reflection coefficients.

Source: Copyright/used with permission of/courtesy of EMW.

simulation using ADS are compared in Figure 3.30. A small mismatch of about 110 MHz, 30 MHz and 80 MHz at first, second and third resonant frequencies, respectively, is observed.

3.5.2 Antenna Design Analysis

The source of each capacitance and inductance in the equivalent circuit is identified. The same values can be calculated using fundamental relations as given in [24]. The inter digital capacitance "C_i" can be calculated using the following equation, as shown in [26]

$$C_i = \frac{\varepsilon_{re} 10^{-3}}{18\pi} \frac{K(k)}{K'(k)}(N-1)l. \tag{3.1}$$

The calculated interdigital capacitance is $C_i = 38$ pF. The self-inductance L of a rectangular strip of length l, width w and thickness t is given by

$$L = 2 \times 10^{-4} l \left[\ln \left(\frac{l}{w+t} \right) + 1.193 + 0.2235 \frac{w+t}{l} \right]. \tag{3.2}$$

Total self-inductance (L_s) is the sum of self-inductance of each section.

$$L_s = \sum_{i=1}^{2N-1} L_i. \tag{3.3}$$

The calculated inductance $L_o = 24.5$ nH, $L_i = 3.5$ nH and $L_t = 12.5$ nH. The capacitances C_g and C_c are calculated by following [24]. The calculated values are $C_g = 92.5$ pF and $C_c = 48$ pF. All other capacitances are calculated by

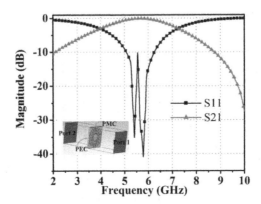

FIGURE 3.31
Simulated return loss and transmission characteristics of MDCsRR.

Source: Copyright/used with permission of/courtesy of EMW.

following [26] and ADS software optimization tool. The equivalent capacitance $C_1 = 400$ nF and $C_2 = 170$ nF.

The geometry of MDCsRR is optimized to get appropriate antenna characteristics. MDCsRR is excited by electric field polarized along the axis of the ring and made to resonate at a frequency determined by equivalent capacitance "C_{eq}" and equivalent inductance "L_{eq}" of the modified ring structure. The resonant frequency of MDCsRR is given by $f_o = 1/2\Pi\sqrt{L_{eq}C_{eq}}$, where

"C_{eq}" is total equivalent capacitance and "L_{eq}" is total equivalent inductance due to MDCsRR. By changing the ring parameters, equivalent inductance and capacitance can be changed and MDCsRR can be made to resonate at a desired frequency. The proposed MDCsRR UC is simulated using CST microwave studio with the appropriate boundary condition, and the simulated return loss and transmission characteristics are plotted in Figure 3.31. It is observed that MDCsRR resonates at two frequencies.

The electric field due to dominant mode propagation within patch cavity is polarized normal to the ground plane. When a modified ring resonator is placed in a time-varying normal magnetic field, an electric field is induced on the metal with a maximum peak value at the resonant frequency. The patch antenna loaded with IDC and MDCsRR is analysed to study its transmission and reflection characteristics. The transmission model of the proposed patch antenna is derived as reported in [22]. The same model is simulated using high-frequency structure simulator CST, and the fabricated prototype is shown in Figure 3.32(a) and (b).

Simulated and measured reflection loss and transmission characteristics of the designed transmission model are compared in Figure 3.33(a) and (b), respectively, which shows good agreement. A small mismatch

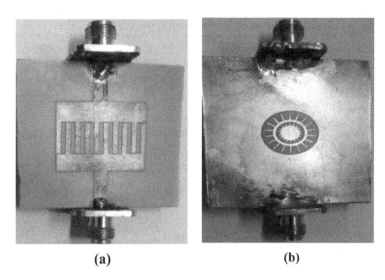

(a) **(b)**

FIGURE 3.32
Fabricated prototype of IDC- and MDCsRR-loaded transmission model; (a) top view and (b) back view.

Source: Copyright/used with permission of/courtesy of EMW.

in measured and simulated results is due to fabrication tolerances. The transmission and reflection parameters predict a band-pass filter characteristic at resonating bands of the proposed antenna. The pass bands with nearly zero reflection indicate zero shunt admittance, and matching of antenna in these bands can thus be possible. In stop bands with zero transmission, matching is not possible due to infinite shunt admittance and hence antenna will not radiate.

The simulated surface current distributions are plotted at resonant sample frequencies in Figure 3.34. Strong surface current concentration can be observed at IDC and MDCsRR. The resonant surface current distribution length is used to validate design strategy for the excitation of triple band as reported in [24]. The same method is applied to validate design strategy of the proposed prototype antenna. The resonating lengths can be calculated using [26] as

$$f_r = \frac{c}{2L_r\sqrt{\varepsilon_{\text{eff}}}} \tag{3.4}$$

$$\varepsilon_{\text{eff}} = \frac{\varepsilon_r + 1}{2} + \frac{\varepsilon_r - 1}{2}\frac{1}{\sqrt{1 + 12\frac{h}{w}}}$$

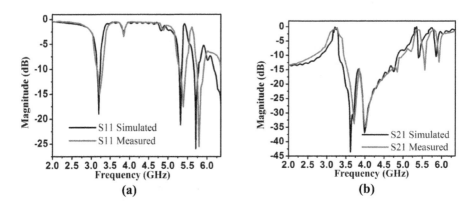

FIGURE 3.33
(a) Reflection characteristics and (b) transmission characteristics of proposed IDC- and MDCsRR-loaded transmission line.

Source: Copyright/used with permission of/courtesy of EMW.

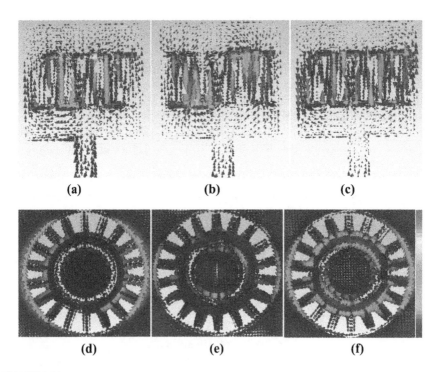

FIGURE 3.34
Surface current distribution on (a) patch at 3.2 GHz; (b) patch at 5.4 GHz; (c) patch at 5.8 GHz; (d) MDCsRR at 3.2 GHz; (e) MDCsRR at 5.4 GHz and (f) MDCsRR at 5.8 GHz.

Source: Copyright/used with permission of/courtesy of EMW.

where first resonance in the proposed antenna is due to central, left- and right-sided highest surface current distribution on IDC finger [Figure 3.34(a)]. At resonance, this length should be half of the wavelength in the medium. Approximate radiating element length responsible for first resonance can be calculated using geometrical dimensions. Resonating length can be calculated as

$$L_{r1} = L_i + W_i + \frac{L_i}{2} + \frac{L_i}{2} + G_i + L_i. \tag{3.5}$$

Using geometrical dimension values, $L_{r1} = 29.80$ mm. The effective dielectric constant (ε_{reff}) is 2.092, calculated using equations given in [27].

At resonance, this length L_{r1} should be $\lambda_g / 2$ and the resonant frequency f_{r1} is given as $f_{r1} = \dfrac{c}{2L_{r1}\sqrt{\varepsilon_{reff}}} \approx 3.48 \text{GHz}$, error generated due to this validation approach is 8 per cent.

The second resonance band is generated due to left- and right-sided surface current distribution on IDC [Figure 3.34(b)]. Approximate length of radiating patch element at this frequency can be calculated as

$$L_{r_2} = \frac{L_i}{2} + G_i + \frac{L_i}{2} + \frac{L_i}{2} + \frac{L_i}{2} + G_i. \tag{3.6}$$

Using geometrical dimension values, $L_{r_2} = 19.20$ mm and at resonance, should be $\lambda_g / 2$ and the resonant frequency f_{r_2} is given as $f_{r_2} = \dfrac{c}{2L_{r_2}\sqrt{\varepsilon_{reff}}} \approx 5.40 \text{ GHz}$; error generated due to this validation approach is 1.3 per cent.

The electrical length contributing to the third resonance band [Figure 3.34 (c)] is given as

$$L_{r_3} = \frac{L_i}{4} + \frac{L_i}{4} + G_i + L_i + \frac{L_i}{2}. \tag{3.7}$$

Using geometrical dimension values, $L_{r_3} = 18.20$ mm and at resonance, should be $\lambda_g / 2$ and the resonant frequency f_{r_3} is given as $f_{r_3} = \dfrac{c}{2L_{r_3}\sqrt{\varepsilon_{reff}}} \approx 5.69 \text{ GHz}$, error generated due to this validation approach is 2.7 per cent.

The surface current distribution plot indicates maxima of current on IDC finger and MDCsRR in the ground plane. Figure 3.35 shows the effect

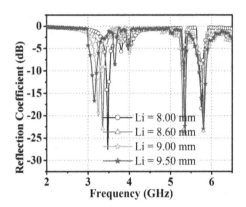

FIGURE 3.35
Reflection coefficient versus frequency for various values of finger length of IDC- "L_i".

Source: Copyright/used with permission of/courtesy of EMW.

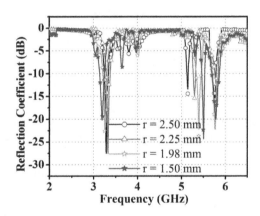

FIGURE 3.36
Reflection coefficient versus frequency for various values of radius of inner circular slot ring resonator "r" of MDCsRR.

Source: Copyright/used with permission of/courtesy of EMW.

of variation of finger length of IDC. IDC finger length "L_i" is changed from 8.00 mm to 9.50 mm. As the length of the finger increases, equivalent capacitance "C_{eq}" also increases, causing resonant frequency "f_o" to decrease as shown in Figure 3.35.

At the second and third resonant bands, surface current is on some part of the IDC, but the main highest current distribution is observed on MDCsRR in the ground plane. For the second resonance band, surface current distribution maxima are mainly concentrated on inner CsRR as shown in Figure 3.34 (e). Figure 3.36 shows effect of variation of radius of inner CsRR on the return loss.

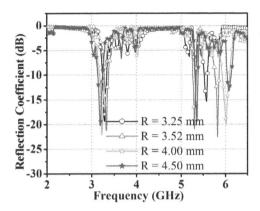

FIGURE 3.37
Reflection coefficient versus frequency for various values of radius of outer circular slot ring resonator "*R*" of MDCsRR.

Source: Copyright/used with permission of/courtesy of EMW.

The radius of inner CsRR "*r*" is changed from 2.50 mm to 1.50 mm. As "*r*" decreases equivalent inductance "L_{eq}" and equivalent capacitance "C_{eq}" decrease, causing increase in the resonant frequency "f_o", as shown in Figure 3.36.

At the third resonant frequency, surface current distribution maximum is mainly concentrated on outer CsRR as shown in Figure 3.34(f). Figure 3.37 shows the effect of variation of radius of outer CsRR. As "*R*" increases, equivalent inductance "L_{eq}" as well as equivalent capacitance "C_{eq}" decreases, causing increase in the resonant frequency "f_o" as shown in Figure 3.37. The effect of change in metallic strip thickness "*t*" of MDCsRR is also studied. No significant change in any of the resonant frequency bands was observed. It is also observed that one can independently control each of the resonant frequencies.

3.5.3 Experimental Results

The proposed patch antenna has electrical size of $0.20\lambda \times 0.20\lambda \times 0.008\lambda$ (19 mm × 19 mm × 0.762 mm), where λ is associated with the first resonance frequency of 3.2 GHz. Conventional patch antenna size operating at 3.2 GHz (31.25 mm × 37.05 mm × 0.762 mm) is miniaturized by 68.83 per cent. The loading of IDC patch antenna with new metamaterial UC causes electrical size reduction as well as multiband operation with significant calculated antenna gain. The top and back views of the fabricated prototype are shown in Figure 3.38(a) and 3.38(b), respectively. The measured and simulated return loss (S_{11}) of the prototype antenna loaded with MDCsRR metamaterial UC shows good agreement as seen from Figure 3.39. The measured 10-dB return loss bandwidth of 80 MHz from 3.17 MHz to 3.25 MHz, 60 MHz from 5.37 MHz to 5.43 MHz and 120 MHz from 5.75 MHz to 5.87 MHz is observed.

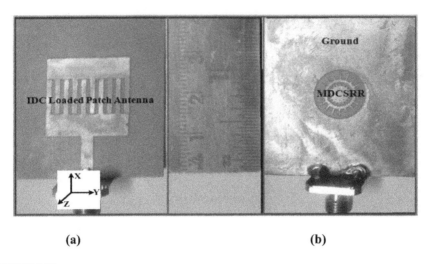

FIGURE 3.38
Fabricated MDCsRR metamaterial UC-loaded microstrip patch antenna; (a) top view) and (b) back view.

Source: Copyright/used with permission of/courtesy of EMW.

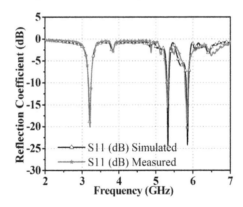

FIGURE 3.39
Simulated and measured reflection coefficient (S_{11}) of the prototype antenna.

Source: Copyright/used with permission of/courtesy of EMW.

The radiation pattern of the fabricated antenna is measured in an anechoic chamber using spectrum analyser and signal generator. Figure 3.40 shows the measured and simulated E- and H-plane radiation patterns, which demonstrate good agreement except for small mismatch in E-plane radiation pattern at 5.4 GHz. The radiation pattern of the antenna with vertical linear electrical field polarization is similar to a short monopole on a finite ground plane. The antenna has both sided radiation pattern due to etched MDCsRR in the ground plane. The back-radiation is less than the forward radiation due

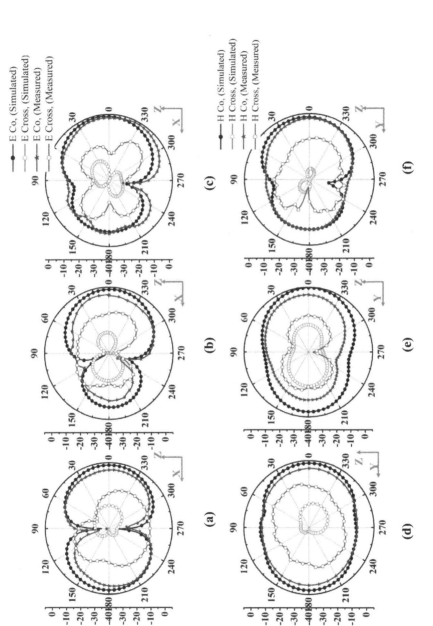

FIGURE 3.40

Radiation pattern of proposed antenna (a) E-plane at 3.2 GHz; (b) E-plane at 5.4 GHz; (c) E-plane at 5.8 GHz; (d) H-plane at 3.2 GHz; (e) H-plane at 5.4 GHz and (f) H-plane at 5.8 GHz.

Source: Copyright/used with permission of/courtesy of EMW.

to the effect of finite ground plane used. Hence proposed antenna has both sided radiation characteristics with higher gain in broadside and medium gain in the backside.

There is good cross-polarization purity at first resonating frequency of 3.2 GHz with a maximum measured electric field cross-polarization level of −16.0 dB in the E-plane and −12.23-dB cross-polarization level in the H-plane. The second band, at resonant frequency of 5.4 GHz, has maximum measured E-plane cross-polarization level of −12.26 dB and H-plane cross-polarization level of −14.21dB.

The third band with a resonant frequency of 5.8 GHz has measured cross-polarization level of −13.38 dB in the H-plane and −22.11 dB in the E-plane. The measured gain of the prototype fabricated antenna as a function of frequency is plotted in Figure 3.41. The measured gain plot indicates significant measured gain in all the three bands with maximum measured gain of 3.28 dBi at first resonant frequency of 3.2 GHz. The second band with resonating frequency of 5.4 GHz and third band with resonating frequency of 5.8 GHz have 2.76 dBi and 3.1 dBi measured gain, respectively. The earlier measured gains are in broadside direction ($\theta = 0°$). The backside ($\theta = 180°$) gain of the proposed antenna is found to be 2.1 dBi, 0.75 dBi and 1.3 dBi at the first, second and third resonant frequencies, respectively. The measured gain of the patch antenna in both sides ($\theta = 0°$ and $\theta = 180°$) is significantly high in all the three bands in spite of small electrical size of the proposed patch antenna. Hence, the proposed antenna is a double-sided radiating patch antenna.

A highly miniaturized significant gain triple-band patch antenna loaded with a new MDCsRR metamaterial UC is successfully demonstrated. The proposed antenna size is miniaturized by about 68.83 per cent as compared to a conventional patch antenna of size 31.25 mm × 37.05 mm × 0.762 mm, operating at first resonant frequency of 3.2 GHz with significant calculated antenna gain. Table 3.2 compares the performance of different multiband antennas reported with the proposed antenna.

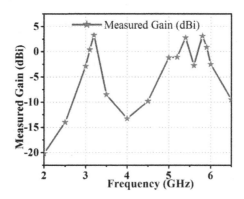

FIGURE 3.41
Measured gain versus frequency plot of the prototype antenna.

Source: Copyright/used with permission of/courtesy of EMW.

TABLE 3.2
Comparison of Proposed Antenna with Other Reported Antennas

Ref.	Size in Terms of λ (wavelength at lowest resonant frequency) X×Y×Z	Frequency Band (GHz)	Calculated Gain (dBi)
[27]	$0.158 \times 0.208 \times 0.013$	2.5, 3.5, 5.5	1.5, 1.7, 3.05
[28]	$0.099 \times 0.149 \times 0.038$	2.4, 5.0	2.1, 5.3
[29]	$0.15 \times 0.15 \times 0.01$	2.6, 3.47, 5.75	0.2, 0.16, 0.62
[30]	$0.13 \times 0.13 \times 0.016$	2.59, 4.73, 5.7	2.59, 3.58, 2.29
Proposed	$0.202 \times 0.202 \times 0.008$	3.2, 5.4, 5.8	3.28, 2.76, 3.1

As observed, the proposed antenna has the smallest size with controlled triple-band operation and offers highest measured broadside and backside antenna gain with 68.83% miniaturization as compared to other reported antennas.

3.6 Conclusion

In this chapter, miniaturization of MPA with multiband operation is proposed. Maximum miniaturization of about 70 per cent is achieved. The fundamental principle to achieve miniaturization is addition of finite capacitance and inductance to increase the series capacitance or inductance and hence reduction in electrical size of the antenna with multiband operation. As a trade-off, increased level of miniaturization causes reduction in −10-dB reflection coefficient bandwidth with good radiation pattern in broadside and backside directions.

In the next chapter, the design of high-gain antennas using reflection-type MS is presented.

References

1. Hansen, R. C. and M. Burke, "Antennas with magneto-dielectrics," *Microwave and Optical Technology Letters*, vol. 26, no. 2, pp. 75–78, 2000.
2. Zhanru, L., X. N. and Z. Jiangao, "New type organic magnetic materials and their application in the field of microwaves," *Journal of Microwares*, vol. 4, pp. 145–170, 1999.

3. Bae, S. and Y. Mano, "A small meander VHF & UHF antenna by magneto-dielectricmaterials," *Asia-Pacific Microwave Conference*, vol. 4, Suzhou, pp. 1–4, 2005.

4. Garg, Ramesh, Prakash Bhartia, Inder J. Bahl, and Apisak Ittipiboon, *Microstrip Antenna Design Handbook*, Artech House, USA, 2001.

5. Shafai, L., "Dielectric loaded antennas," in *Wiley Encyclopedia of Electrical and Electronics Engineering*, USA, 1999.

6. Volakis, J., C.-C. Chen, and K. Fujimoto, *Small Antennas: Miniaturization Techniquesand Applications*, McGraw-Hill Companies, New York, 2010.

7. Kang, C.-H., S.-J. Wu, and J.-H. Tarng, "A novel folded UWB antenna for wireless body area network," *IEEE Transactions on Antennas and Propagation*, vol. 60, no. 2, pp. 1139–1142, 2012.

8. Scardelletti, M., G. Ponchak, S. Merritt, J. Minor, and C. Zorman, "Electrically small folded slot antenna utilizing capacitive loaded slot lines," *Radio and Wireless Symposium, 2008 IEEE*, pp. 731–734, 2008.

9. Chi, P.-L., K. Leong, R. Waterhouse, and T. Itoh, "A miniaturized CPW-fed capacitor loaded slot-loop antenna," *Signals, Systems and Electronics, 2007. ISSSE'07. International Symposium on Signals, Systems and Electronics*, Montreal, pp. 595–598, 2007.

10. Lee, D. H., A. Chauraya, Y. Vardaxoglou, and W. S. Park, "A compact and low-profile tunable loop antenna integrated with inductors," *IEEE Antennas and Wireless Propagation Letters*, vol. 7, pp. 621–624, 2008.

11. Sievenpiper, D., L. Zhang, R. Broas, N. Alexopolous, and E. Yablonovitch, "Highimpedance electromagnetic surfaces with a forbidden frequency band," *IEEE Transactions on Microwave Theory and Techniques*, vol. 47, no. 11, pp. 2059–2074, 1999.

12. Yang, F. and Y. Rahmat-Samii, "Reflection phase characterizations of the EBG ground plane for low profile wire antenna applications," *IEEE Transactions on Antennas and Propagation*, vol. 51, no. 10, pp. 2691–2703, 2003.

13. Yang, F. and Y. Rahmat-Samii, "Reflection phase characterizations of the EBG ground plane for low profile wire antenna applications," *IEEE Transactions on Antennas and Propagation*, vol. 51, no. 10, pp. 2691–2703, 2003.

14. Maagt, P., R. Gonzalo, Y. C. Vardaxoglou, and J. M. Baracco, "Electromagnetic band gap antennas and components for microwave and (sub)millimeter wave applications," *IEEE Transactions on Antennas and Propagation*, vol. 51, no. 10, pp. 2667–2677, 2003.

15. Caloz, C., T. Itoh, and A. Rennings, "CRLH metamaterial leaky-wave and resonant antennas," *IEEE Antennas and Propagation Magazine*, vol. 50, no. 5, 2008.

16. Stockman, M. I., "Criterion for negative refraction with low optical losses from a fundamental principle of causality," *Physical Review Letters*, vol. 98, no. 17, p. 177404, 2007.

17. Singh, A. K., Mahesh P. Abegaonkar, and Shiban K. Koul, "Highly miniaturized dual band patch antenna loaded with metamaterial unit cell," *Microwave and Optical Technology Letters*, vol. 59, no. 8, pp. 2027–2033, May 2017.

18. Ha, Jaegeun, Kyeol Kwon, Youngki Lee, and Jaehoon Choi, "Hybrid mode wideband patch antenna loaded with a planar metamaterial unit cell," *IEEE Transactions on Antennas and Propagation*, vol. 60, no. 2, pp. 1143–1147, 2012.

19. Guterman, Jerzy, A. A. Moreira, and C. Peixerio, "Dual-band miniaturized microstrip fractal antenna for a small GSM 1800+ UMTS mobile handset," *Electrotechnical Conference, 2004. MELECON 2004. Proceedings of the 12th IEEE Mediterranean*, vol. 2, IEEE, Croatia, pp. 499–501, 2004.
20. Al-Joumayly, Mudar A., Suzette M. Aguilar, Nader Behdad, and Susan C. Hagness, "Dual-band miniaturized patch antennas for microwave breast imaging," *IEEE Antennas and Wireless Propagation Letters*, vol. 9, pp. 268–271, 2010.
21. Zhu, J. and G. V. Eleftheriades, "Dual-band metamaterial-inspired small monopole antenna for Wi-Fi applications," *Electronics Letters*, vol. 45, no. 22, pp. 1104–1106, Oct. 22, 2009.
22. Singh, Amit Kumar and P. R. Chadha, "U shaped multiband microstrip patch antenna for wireless communication system and parametric variational analysis," *Wireless and Optical Communications Networks WOCN2013*, Bhopal, pp. 1–4, 2013.
23. Singh, Amit K., Mahesh P. Abegaonkar, and Shiban K. Koul, "Triple band miniaturized patch antenna loaded with metamaterial unit cell for defense applications," *2016 11th International Conference on Industrial and Information Systems (ICIIS)*, Roorkee, pp. 833–837, 2016.
24. Antoniades, M. A. and G. V. Eleftheriades, "Multiband compact printed dipole antennas using NRI-TL metamaterial loading," *IEEE Transactions on Antennas and Propagation*, vol. 60, no. 12, pp. 5613–5626, Dec. 2012.
25. Singh, Amit K., Mahesh P. Abegaonkar, and Shiban K. Koul, "Miniaturized multiband microstrip patch antenna using metamaterial loading for wireless application," *Progress in Electromagnetics Research C*, vol. 83, 71–82, March 2018.
26. Kurra, Lalithendra, Mahesh P. Abegaonkar, and Shiban K. Koul, "Equivalent circuit model of resonant EBG band stop filter," *IETE Journal of Research*, vol. 62, no. 1, pp 17–26, Jan. 2016.
27. Gautam, A. K., L. Kumar, B. K. Kanaujia, and K. Rambabu, "Design of compact F-shaped slot triple-band antenna for WLAN/WiMAX applications," *IEEE Transactions on Antennas and Propagation*, vol. 64, no. 3, pp. 1101–1105, March 2016.
28. Brocker, D. E., Z. H. Jiang, M. D. Gregory, and D. H. Werner, "Miniaturized dual-band folded patch antenna with independent band control utilizing an interdigitated slot loading," *IEEE Transactions on Antennas and Propagation*, vol. 65, no. 1, pp. 380–384, Jan. 2017.
29. Boukarkar, A., X. Q. Lin, Y. Jiang, and Y. Q. Yu, "Miniaturized single-feed multiband patch antennas," *IEEE Transactions on Antennas and Propagation*, vol. 65, no. 2, pp. 850–854, Feb. 2017.
30. Ali, T. and R. C. Biradar, "A triple band highly miniaturized antenna for WiMAX/WLAN applications," *Microwave and Optical Technology Letters*, vol. 60, pp. 466–471, 2018.

4

High-Gain Antennas Using a Reflection-Type Metasurface

4.1 Introduction

Fabry–Perot cavity (FPC) antennas are a type of highly directive planar antenna that offer a promising alternative to standard planar microstrip patch arrays or WG slot array antennas. They offer significant advantages in terms of low fabrication complexity, high radiation efficiency and good radiation pattern performance. These advantages, in conjunction with a renewed interest in periodic surfaces and MSs, led to reinvigoration of international research on this antenna type. This chapter reports recent advances on the design and implementation of FPC antennas for X-band and C-band applications. The main concept of FPAs, their operating principles and analysis approaches are briefly introduced. Highly reflective MS is designed first and experimentally characterized. The reflection-type MS is used as a partially reflecting surface (PRS) for designing open-air FPC resonators. The main FPC antenna structure is formed by a PRS placed at a distance of about a half-wavelength and in parallel to a ground plane, thus forming an open Fabry–Perot-type resonant cavity. An excitation source is typically included within the cavity, e.g. a dipole, a microstrip patch or a slot in the ground plane. The gain and bandwidth of the FPC antenna depend on reflection (amplitude and phase) from the PRS as well as the distance from the ground plane. The PRS can be either a passive periodic array, such as FSSs, which are customarily used for filtering EM waves, or an MS with sub-wavelength and potentially nonuniform UCs. Several FPC antennas are reported, in [1–15], to enhance the radiation characteristic of microwave source radiator. In this chapter, a low-profile highly directive FPC antenna using highly reflective MS is designed for C- and X-band applications. These FPC antennas can also be called leaky wave antennas or PRS antennas.

The FPC can be designed by placing highly reflective MS on top of a ground plane, forming an open-air cavity. The open-air cavity is excited by placing a microwave source like an MPA in the ground plane. The height of MS over the ground plane can be optimized to generate constructive

DOI: 10.1201/9781003045885-4

interference. The cavity height and MS can control overall gain and bandwidth of the cavity radiator. The reflecting surface can be designed using uniform or nonuniform arrangement of fundamental UC, single or multiple layers of dielectric of same or different dielectric constant and single or multiple layers of highly reflective MS. Metal FPC, dielectric FPC and metal–dielectric (metal with dielectric) FPC are several types of FPC antennas depending on the type of reflecting surface. The working principle of an FPC antenna is explained by using a ray-tracing model in Figure 4.1. The ray-tracing model is proposed by considering the reflecting surface as an infinite arrangement, thus not considering the diffraction effect and ignoring higher-order modes.

4.2 Working Principle

A ray-tracing model of the proposed structure due to multiple reflections inside the cavity and leaked-out waves from metamaterial superstrate is shown in Figure 4.1(a). A conducting ground plane behind the patch antenna acts as a shield against backward radiation. Ground plane and reflecting metamaterial superstrate act as parallel plate, where multiple reflections take place in between. Radiated waves from the patch antenna strike on metamaterial superstrate at different angles. An MS reflects waves depending on the reflection coefficient (magnitude and phase) of reflecting surface and wave incidence angle "α" as shown in Figure 4.1(b). Total internal reflection of waves occurs when incidence angle is greater than the critical angle. As the angle of incidence increases, reflection by MS inside FPC decreases. Due to this, strong radiation from the centre of MS and weaker towards margin sides are observed. Wave propagation through dielectric occurs as reported in [15]. Multiple reflections inside the cavity cause increase in antenna gain in the direction of boresight ($\theta = 0°$). Leaking out of waves in the normal direction from reflecting MS causes an increase in radiation property of MPA.

The cavity height is selected typically close to half of resonating wavelength to achieve constructive interference for waves bouncing between MS and ground plane.

Air gap "h" between the ground plane and reflecting MS is calculated such that leaked-out waves through the reflecting surface outside the cavity have equal phase in normal direction. The proposed MPA with MS, forming a cavity resonator, resonates at a frequency given by $H \cong N\lambda / 2$. Here "H" is the distance between the ground plane and reflecting MS superstrate [Figure 4.1(b)] and is given by $H = h + t + d$ and "N" is any integer value like 1, 2, 3... Consider MS placed on the antenna at a height "h" from the antenna surface and take two points P_1 and P_2 on the MS separated by a distance of "a".

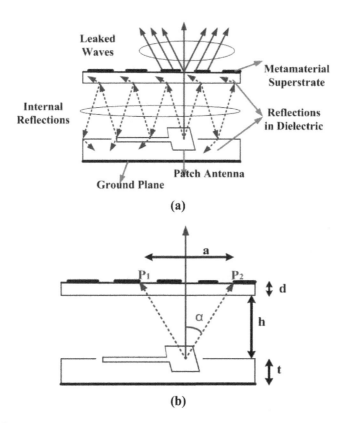

FIGURE 4.1
(a) Ray-tracing model of proposed cavity antenna and (b) various parameters of the cavity resonator.

Source: Copyright/used with permission of/courtesy of IEEE.

Further, consider patch antenna acting as a radiating point source. The electric field at points P_1 and P_2 will have a phase difference of " $\Delta\ \Phi$ ", where " $\Delta\ \Phi$ " can be given as

$$\Delta\Phi = \beta o\left(\sqrt{h^2 + a^2} - h\right) - 2n\pi .$$ (4.1)

Here, in Equation 4.1 " βo " is the propagation constant. The FPC is excited by patch antenna inside and is coupled to free space with front side of the MS. Gain of the antenna can be increased by controlling higher-order modes propagating through the cavity. The number of modes excited in the FPC depends on location of MPA and cavity dimensions. The directivity performance of an FPC antenna can be realized by characterizing the phase response of the reflecting surface under normal incidence of EM waves. The complex wave number of leaky wave modes propagating inside MS can predict the direction and beam width of the main radiated beam.

Reduced profile FPC antenna can also be designed using a reflecting MS with reflection zero phase at some frequency acting as an AMC. The zero-reflection phase can change the resonance condition of the cavity. The ground plane of FPC antenna can also be changed by AMC causing change in resonance condition and hence reduced FPC antenna profile.

4.3 Design of a High-Gain and High Aperture Efficiency Cavity Resonator Antenna for X-Band Applications Using a Reflection-Type Metamaterial Superstrate

In this section, a high-gain cavity resonator antenna using MPA and a metamaterial surface superstrate is designed. A low-profile highly reflective MS with wideband reflection characteristic is designed first. Next, a cavity is designed using MPA and designed MS, forming an FPC. The designed FPC resonator antenna is characterized experimentally. The aperture of the designed FPC radiator is found to be highly efficient [16].

4.3.1 Design of an FPC Resonator Antenna

The highly reflective MS proposed in Chapter 2, Section 2.4, is used as a reflecting surface on top of a perfectly ground-backed planar patch antenna acting as a microwave radiator. The designed array of DASR UC MS (Section 2.4) is a wideband highly reflective surface with 3-dB reflection bandwidth of about 4.10 GHz from 8.1 GHz to 12.2 GHz. This MS has a maximum reflection of −45 dB at a frequency of 10.10 GHz. The fabricated prototype of MS is shown in Figure 4.2(a). This MS is placed on top of a ground-backed microwave radiator to form an open-air cavity. The MS as a superstrate on the patch antenna radiator acts as an FPC. The proposed patch antenna is a microstrip-fed rectangular patch antenna fabricated on Neltec substrate having relative permittivity of 2.2 and a thickness of 0.762 mm, resonating at 10.09 GHz with calculated antenna gain of 4.5 dB. Patch antenna of size 10 mm × 10 mm is fabricated on a ground plane of 65 mm × 65 mm as shown in Figure 4.2(b) and (c).

FPC is a resonant cavity whose resonant frequency depends on optimum cavity length. The cavity length decides the propagating mode inside the cavity due to multiple reflections. Cavity length "h" is varied from 14 mm to 20 mm and return loss response of the cavity is plotted in Figure 4.3. The optimum cavity length is found to be 15.5 mm and is used in the design process. The prototype FPC antenna is shown in Figure 4.4, where the designed array of DASR UC acting as reflecting MS is on the top of the MPA radiator.

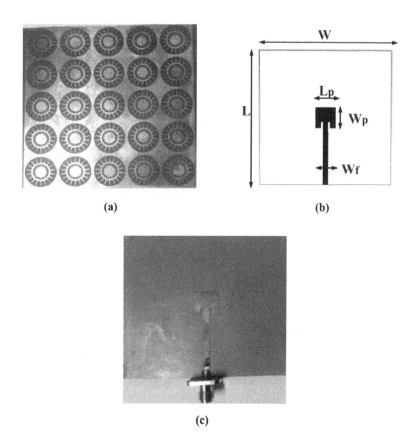

(a) **(b)**

(c)

FIGURE 4.2
(a) Fabricated prototype metasurface superstrate; (b) microstrip patch antenna geometry and (c) fabricated prototype patch antenna. Dimensions are $L = 65$, $W = 65$, $L_p = 10$, $W_p = 10$, $W_f = 2.50$ (all dimensions are in millimetres).

Source: Copyright/used with permission of/courtesy of IEEE.

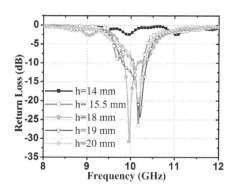

FIGURE 4.3
Return loss of proposed antenna for various cavity lengths.

Source: Copyright/used with permission of/courtesy of IEEE.

FIGURE 4.4
Fabricated metamaterial superstrate Fabry–Perot cavity resonator antenna.

Source: Copyright/used with permission of/courtesy of IEEE.

4.3.2 Measured Results of an FPC Antenna

The return loss characteristics of the proposed cavity resonator antenna and the patch antenna are discussed by comparing their simulated and measured results. The return loss characteristics of MPA and FPC antenna are measured using VNA. As seen from Figure 4.5, there is good agreement between simulated and measured return loss characteristics in both cases. A small mismatch in measured and simulated results is observed. This small mismatch is due to the loading effect of MS when placed on top of an MPA radiator. The reflected waves from MS can cause multiple reflections inside the cavity, resulting in small mismatch in return loss measurement.

The radiation characteristics of both patch and FPC patch antennas are measured in an anechoic chamber using spectrum analyser and signal generator. The realized gain pattern of the antennas for vertical linear electric field polarization and magnetic field polarization with simulated and measured results are compared in Figure 4.6 and Figure 4.7, respectively.

As observed, the simulated and measured results are in good agreement. As compared to MPA, calculated gain enhancement of 11.85 dB in H-plane and 12.5 dB in E-plane is observed at 10.18 GHz. The broadside gain of FPC antenna is 16.35 dBi. Measured results also indicate that a good cross-polarization level is observed in H-plane as well as in E-plane.

The simulated and measured antenna gain of both patch and FPC patch antennas in H-plane at various frequencies from 8 GHz to 12 GHz is measured and compared in Figure 4.8. It is seen that gain of patch antenna is 4.5 dBi and that of FPC patch antenna is 16.35 dBi. Hence, gain enhancement of 11.85 dB is observed. Antenna aperture efficiency of MPA and FPC antenna is calculated as discussed in [3].

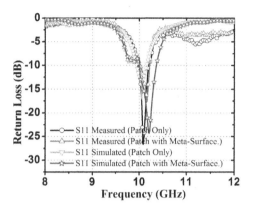

FIGURE 4.5
Measured and simulated return loss of patch antenna with and without superstrate.

Source: Copyright/used with permission of/courtesy of IEEE.

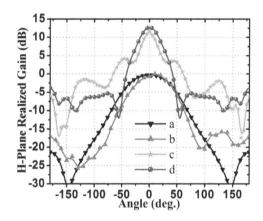

FIGURE 4.6
Realized gain in H-plane (a) patch antenna simulated; (b) patch antenna measured; (c) patch antenna with superstrate measured and (d) patch antenna with superstrate simulated.

Source: Copyright/used with permission of/courtesy of IEEE.

$$G_{max} = \left(\tfrac{4\pi}{\lambda_o^2}\right) A_{ef} E_f. \tag{4.2}$$

Here, in Equation 4.2, A_{ef} is effective aperture area, E_f is radiation efficiency and G_{max} is maximum gain due to effective aperture area. The calculated patch antenna gain is 25.84 per cent of maximum gain (G_{max}) due to effective aperture area (A_{ef} = 65 mm × 65 mm), whereas calculated FPC antenna gain is 92.42 per cent of G_{max} due to A_{ef}. Hence, the proposed MS aperture is highly efficient. For effective radiating aperture of 65 mm × 65 mm, G_{max}

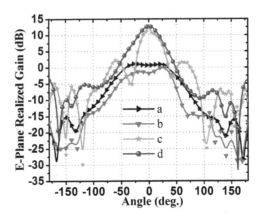

FIGURE 4.7
Realized gain in E-plane (a) patch antenna simulated; (b) patch antenna measured; (c) patch antenna with superstrate measured and (d) patch antenna with superstrate simulated.

Source: Copyright/used with permission of/courtesy of IEEE.

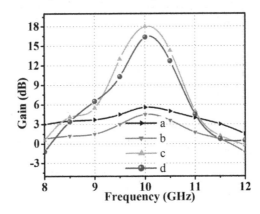

FIGURE 4.8
Gain of antenna in H-plane (a) patch antenna simulated; (b) patch antenna measured; (c) patch antenna with superstrate simulated and (d) patch antenna with superstrate measured.

Source: Copyright/used with permission of/courtesy of IEEE.

for original patch antenna is 17.41 dB whereas for FPC patch antenna, G_{max} is 17.69 dB. The measured gain of original patch antenna is 25.84 per cent of its G_{max} whereas the measured gain of FPC patch antenna is 92.42 per cent of its G_{max}. Hence, aperture efficiency of FPC patch antenna is much higher than aperture efficiency of original patch antenna.

A high-gain cavity resonator antenna using MS superstrate is successfully demonstrated. It is observed that the proposed MS superstrate causes gain enhancement of 11.85 dB in H-plane and 12.5 dB in E-plane at 10.18 GHz

TABLE 4.1

Comparison of the Proposed Antenna with Other Reported Antennas

Ref.	Substrate (ε_r)	Substrate Thickness (mm)	Distance of Superstrate from Patch Antenna	Gain Enhancement (dB)	Calculated Aperture Efficiency (in terms of G_{max})	Type of Superstrate and Number of Superstrates
[10]	10	10	$0.386\lambda_o$	2.3	57.60%	Transmission, 1
[11]	3.55	0.80	$0.813\lambda_o$	7.8	70.00%	Transmission, 3
[12]	3.2	0.762	$0.522\lambda_o$	6.95	83.00%	Transmission, 1
[13]	3.2	0.762	$0.51\lambda_o$	4.5	65.20%	Transmission, 2
[14]	4.4	1.60	$0.60\lambda_o$	2.65	68.70%	Reflection, 1
[15]	3.48	0.762	$0.27\lambda_o$	3.48		Reflection, 3
[17]	3.48	0.762	$0.555\lambda_o$	6.2	80.00%	Reflection, 1
This Work	2.2	0.762	$0.578\lambda_o$	12.5	92.42%	Reflection, 1

as compared to conventional MPA, with improved front-to-back ratio (FBR) and cross-polarization level. The proposed MS aperture is highly efficient, achieving 92.42 per cent of G_{max}.

The proposed antenna is compared with several similar reported —antennas in Table 4.1. As observed, the proposed FPC patch antenna is found to have the highest aperture efficiency. The proposed antenna also provides the highest gain enhancement with less complexity (number of superstrates).

4.4 Wideband Gain Enhancement of an FPC Antenna Using a Reflecting Metasurface for C-Band Applications

In this section, both narrow and wideband high-gain FPC antennas for C-band applications with improved FBR using highly reflective metamaterial surfaces are proposed [18]. A highly reflective compact ultrathin MS for C-band application is designed first and a cavity is created by placing the MS on the top of the radiator as reported in [19]. The designed FPC antennas are experimentally characterized. This antenna is designed for 5G sub 6-GHz band applications. The antenna is found to have high-gain, wide-gain enhanced bandwidth and high aperture efficiency.

4.4.1 Design of a Highly Reflective Metasurface

A low-profile maximum-reflection MS operating at 5.3 GHz with wide reflection band is designed first. The proposed MS is designed by modifying a CRR. A simple CRR is designed first. Next, to get the optimum resonance frequency band and to achieve compactness, capacitive slots are introduced on the metallic ring, due to which the overall electrical dimensions are reduced and a compactness of 25 per cent, as compared to other conventional CRRs operating in the same resonant band, is achieved. The designed UC is modified circular ring resonator (MCRR). The MCRR UC with detailed dimensional geometry, shown in Figure 4.9, is simulated with PEC along the X-axis and PMC along the Y-axis and WG ports along Z-axis. The MCRR is simulated using CST microwave studio software. The simulated transmission and reflection magnitude response with respective phase response is plotted in Figure 4.10(a), (b), respectively. A wide –10-dB reflection band of 1.35 GHz, indicating more than 90 per cent reflection, is obtained with centre frequency of 5.3 GHz. The material parameters of MCRR are extracted using standard free space measurement technique as discussed in Chapter2. The extracted refractive index of the MCRR UC is given in Figure 4.10(c) and found to be negative in the range of the desired operating band.

The surface current distribution at centre reflection band frequency of 5.3 GHz is studied and found to be maximum on the capacitive slot dimensions and about the MCRR mean radius "R" as shown in Figure 4.11(a). The surface current density is not uniform on the surface of the UC.

Maximum surface current is concentrated on the middle region, whereas from the middle region to the inner side as well as the outer side, the surface current density decreases. The MCRR mean radius "R" and capacitive slot thickness "t" can control the resonant pass- and stop-bands. To analyse this, dimensions "R" and "t" are varied and changes in reflection coefficient are observed. The variations in transmission and reflection characteristics with changing "R" and "t" are plotted in Figure 4.11 (b) and Figure 4.11 (c), respectively. It is observed that increasing "R" decreases the stop-band centre frequency due to increase in effective inductance and capacitance value and vice versa. As the capacitive slot length "t" increases, the effective capacitance increases, resulting in decrease in the stop-band centre frequency.

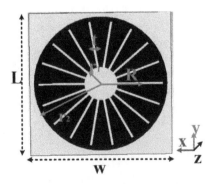

FIGURE 4.9

The geometry of the proposed MCRR UC with dimension. The dimensions are $L = 14$ mm, $W = 14$ mm, $r_1 = 2$ mm, $r_2 = 8$ mm, $R = 5$ mm and $t = 0.15$ mm.

Source: Copyright/used with permission of/courtesy of Wiley.

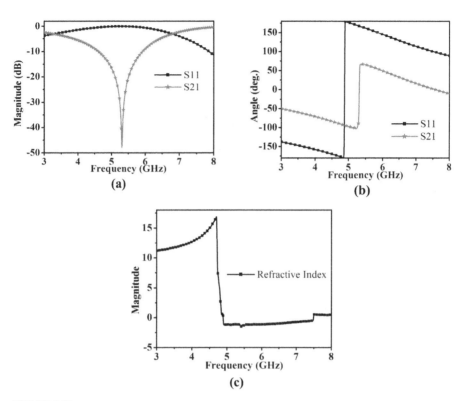

FIGURE 4.10
The simulated transmission and reflection characteristics (a) magnitude plot (b) phase plot and (c) the extracted refractive index.

Source: Copyright/used with permission of/courtesy of Wiley.

Maximum reflection MS is designed using 5 × 5 array of proposed MCRR. The MS is fabricated using photolithography on RT/duroid 5880 substrate having permittivity 2.22 and thickness of 0.762 mm.

The fabricated prototype of MS is shown in Figure 4.12. The thickness of MS is 0.0134λ (0.762 mm), where λ is the wavelength at resonant frequency of 5.3 GHz, making the designed MS ultrathin and highly reflective.

4.4.2 Design of the Narrow-Band FPC Antenna and Working Principle

To design a high-gain, low-profile narrow-band FPC antenna, a cavity is created using a standard narrow-band microwave radiator with an extended ground plane, and the designed MS is placed over the radiator. The extended ground plane acts as PEC and maximum reflection MS over the radiator forms an FPC. The FPC designed earlier is considered as a low-profile high-gain FPC antenna. A standard extended-ground narrow-band patch antenna acts as a radiator. Proper excitation of the cavity depends on two parameters, position of radiator and the height of the cavity.

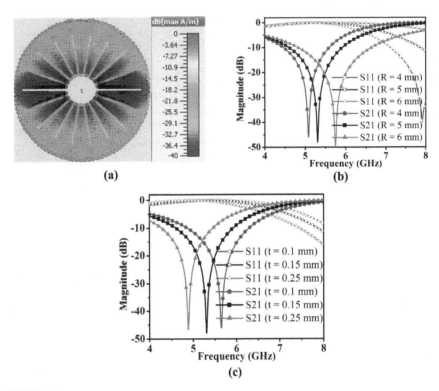

(a)

(b)

(c)

FIGURE 4.11

(a) Simulated surface current distribution on MCRR; the simulated transmission and reflection characteristics (b) with variation of "*R*" and (c) with variation of "*t*".

Source: Copyright/used with permission of/courtesy of Wiley.

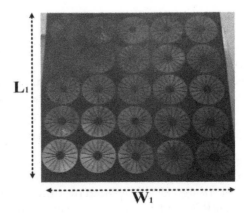

FIGURE 4.12

The prototype of fabricated highly reflective C-band MS. The dimensions are L_1 = 70 mm and W_1 = 70 mm.

Source: Copyright/used with permission of/courtesy of Wiley.

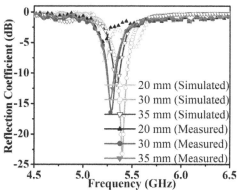

FIGURE 4.13

The measured and simulated reflection coefficients with variation of cavity height "*h*".

Source: Copyright/used with permission of/courtesy of Wiley.

The patch antenna radiator position and the height of cavity are optimized using CST microwave studio to get proper matching condition of FPC, resulting in performance improvement. The variation in measured, and the simulated reflection coefficient due to variation of cavity height "*h*" is plotted in Figure 4.13. The optimum cavity height "$h_{opt} = 30$ mm" is selected to provide maximum matching and perfect excitation of FPC. The resonance frequency of FPC resonator antenna can be obtained using Equation 4.3.

$$\Phi_{MS} + \Phi_{G} = \frac{4\pi h_{opt}}{c} f + 2n\pi \tag{4.3}$$

Considering the first resonance of the FPC ($n = 0$), "Φ_{G}" is reflection phase due to ground plane and "Φ_{MS}" is the reflection phase due to MS. From the Equation 4.3, $h_{opt} \cong \frac{\lambda}{2}$. The selected optimum height value "$h_{opt} = 30$ mm" satisfies Equation 4.3. Maximum directivity of FPC antenna is given by [20]

$$D = 10\log\frac{1+|\Gamma|}{1-|\Gamma|} \tag{4.4}$$

Here, "Γ" is the reflection coefficient magnitude of MS. highly reflective MS will cause maximum directivity. The designed MS has simulated "$\Gamma = 0.997$", indicating maximum directivity of 16.0 dB. The waves leaking out from MS after multiple reflections inside the cavity cause gain enhancement in the broadside direction as discussed in [17].

4.4.3 Measured Results of a Narrow-Band FPC Antenna

The designed prototype of the high-gain narrow-band FPC antenna with the proposed MCRR MS placed over the patch antenna radiator at an optimum height of "$h_{opt} = 30$ mm" is shown in Figure 4.14. The MPA and reflection-type MS are fabricated on RT/duroid 5880 substrate having permittivity 2.2 and thickness of 0.762 mm. Spacers are used to provide optimum cavity height for designing the FPC resonator antenna.

The reflection coefficients of patch antenna and FPC antenna are measured using VNA. The simulated and measured reflection coefficients of FPC antenna and MPA are plotted in Figure 4.15. As observed, the measured resonant frequency is shifted by 0.12 GHz as compared to the simulated values. The earlier error is due to finite dimension of MS and effect due to loading of patch antenna by reflecting MS.

The radiation characteristic of FPC and patch antennas is measured in an anechoic chamber. The measured and simulated H-plane radiation characteristics of MPA and FPC antennas are plotted in Figure 4.16. A good

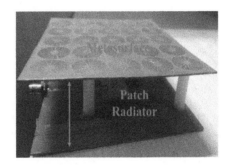

FIGURE 4.14
The narrow-band high-gain FPC antenna prototype.

Source: Copyright/used with permission of/courtesy of Wiley.

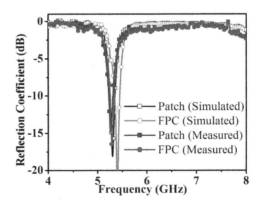

FIGURE 4.15
The measured and simulated reflection coefficients of FPC and patch antennas.

Source: Copyright/used with permission of/courtesy of Wiley.

agreement between simulated and measured results is observed. Broadside gain enhancement of 8.74 dB with 30° reductions in half-power beam width (HPBW) is obtained. The E-plane radiation characteristic is measured and plotted in Figure 4.17. Again, enhancement of 8.50 dB with 35° reductions in HPBW is observed in E-plane in the broadside direction. The reduction in HPBW causes beam focusing, resulting in gain enhancement. It is observed that the gain enhancement of FPC antenna can be controlled by varying optimum cavity height. The variation of cavity height changes the cavity

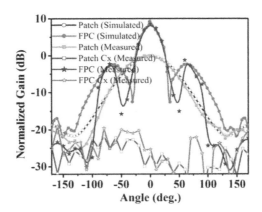

FIGURE 4.16
The measured and simulated radiation patterns of FPC and patch antennas in H-plane. Co and cross-polarized (Cx).

Source: Copyright/used with permission of/courtesy of Wiley.

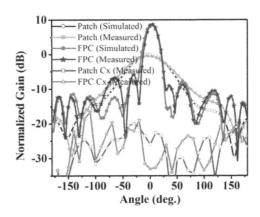

FIGURE 4.17
The measured and simulated radiation patterns of FPC and patch antenna in E-plane. Co and cross-polarized (Cx).

Source: Copyright/used with permission of/courtesy of Wiley.

resonance conditions, resulting in broadside gain variation. It is observed that the designed FPC antenna has narrow enhanced gain bandwidth. The measured cross-polarization level in broadside direction is improved by about 10 dB as shown in Figures 4.16 and 4.17.

4.4.4. Wideband FPC Antenna Design and Measured Results

A wideband FPC antenna is designed by placing the same fabricated highly reflective ultrathin MCRR MS over a standard wide C-band WG radiator with extended ground plane as shown in Figure 4.18. The MCRR MS is placed over the radiator at optimum height "h_{opt} = 30 mm" so that the FPC matching condition "$h_{opt} \cong \dfrac{\lambda}{2}$" can be achieved and cavity can be properly excited at a resonant frequency of 5.30 GHz. The reflection coefficient is measured using VNA. The measured reflection coefficient results predict that the wideband FPC antenna prototype has maximum reflection dip at 5.30 GHz, indicating maximum resonance at 5.30 GHz.

The radiation characteristics of standard wide C-band WG radiator and designed wideband FPC antenna are measured in an anechoic chamber. The measured H-plane radiation characteristics of both the antennas at maximum reflection frequency of 5.30 GHz are compared in Figure 4.19. A broadside gain enhancement of 8.32 dB is observed.

The H-plane radiation characteristics of both antennas are measured at several different frequencies from 5 GHz to 8 GHz. The optimum height of MS over WG radiator is analysed at different frequencies, and the measured broadside gain of WG FPC antenna at their optimum height is plotted with frequency in Figure 4.20. A wide-gain enhancement band of more than 1 GHz from 4.75 GHz to about 6.00 GHz is observed. It is observed that the controlled gain variation can be obtained by varying the optimum cavity height. The measured FBRs of both antennas from 5 GHz to 8 GHz are compared in Figure 4.21. A wide FBR improvement band of 1 GHz from 4.80 GHz to 5.80 GHz is observed.

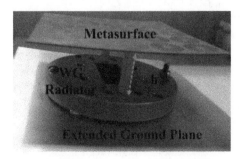

FIGURE 4.18
Photograph of the designed prototype wideband FPC antenna.

Source: Copyright/used with permission of/courtesy of Wiley.

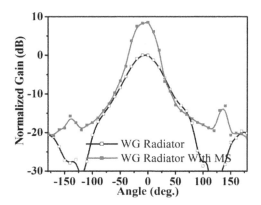

FIGURE 4.19
The measured radiation pattern of WG radiator and wideband FPC antenna at 5.3 GHz.

Source: Copyright/used with permission of/courtesy of Wiley.

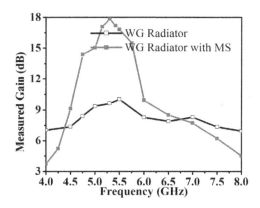

FIGURE 4.20
Measured gain of WG radiator and wideband FPC antenna.

Source: Copyright/used with permission of/courtesy of Wiley.

The design of narrow-band and wideband gain enhanced FPC antenna is presented in this chapter. Next, a narrow-band FPC antenna with broadside gain enhancement of 8.54 dB having maximum measured gain of 16.5 dB is designed at 5.3 GHz. A wideband FPC antenna is designed and wide enhanced gain bandwidth of about 1.00 GHz from 4.75 GHz to 6.00 GHz with maximum gain enhancement of 8.32 dB at 5.30 GHz is obtained.

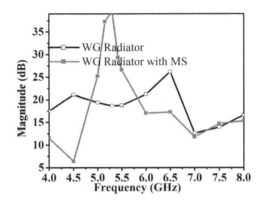

FIGURE 4.21
Measured front-to-back ratio (FBR) of WG radiator and wideband FPC antenna.

Source: Copyright/used with permission of/courtesy of Wiley.

4.5 Conclusion

In this chapter, the design of an FPC resonator antenna using low-profile highly reflective MS for C-band and X-band applications is successfully demonstrated. It is observed that the reflecting MS and the cavity height plays an important role in the design of an FPC antenna. The FPC resonator antenna is found to be a highly directive antenna where the enhanced gain bandwidth can be controlled using wideband radiator. The maximum gain enhancement of 12.5 dB and 8.54 dB in X-band and C-band applications, respectively, is observed.

In the next chapter, the design and development of high-gain antennas for C-band and X-band applications using low-profile transmission-type MSs are presented.

References

1. Chen, L., Z. Y. Lei, R. Yang, J. Fan, and X. W. Shi, "A broadband artificial material for gain enhancement of antipodal tapered slot antenna," *IEEE Transactions on Antennas and Propagation*, vol. 63, no. 1, pp. 395–400, Jan. 2015.
2. Ji, L. Y., Y. J. Guo, P. Y. Qin, S. X. Gong, and R. Mittra, "A reconfigurable partially reflective surface (PRS) antenna for beam steering," *IEEE Transactions on Antennas and Propagation*, vol. 63, no. 6, pp. 2387–2395, June 2015.
3. Liu, Z. G., Z. X. Cao, and L. N. Wu, "Compact low-profile circularly polarized Fabry–Perot resonator antenna fed by linearly polarized microstrip patch," *IEEE Antennas and Wireless Propagation Letters*, vol. 15, pp. 524–527, 2016.

4. Xie, P., G. Wang, T. Cai, H. Li, and J. Liang, "Novel Fabry-Perot cavity antenna with enhanced beam steering property using reconfigurable meta-surface," *Applied Physics A*, vol. 123, no. 7, p. 462, 2017.

5. Xie, P. and G.-M. Wang, "Design of a frequency reconfigurable Fabry-Perot cavity antenna with single layer partially reflecting surface progress," *Electromagnetics Research Letters*, vol. 70, pp. 115–121, 2017.

6. Qin, F. et al., "A triband low-profile high-gain planar antenna using Fabry–Perot cavity," *IEEE Transactions on Antennas and Propagation*, vol. 65, no. 5, pp. 2683–2688, May 2017.

7. Ratni, B., W. A. Merzouk, A. de Lustrac, S. Villers, G. P. Piau, and S. N. Burokur, "Design of phase-modulated metasurfaces for beam steering in Fabry–Perot cavity antennas," *IEEE Antennas and Wireless Propagation Letters*, vol. 16, pp. 1401–1404, 2017.

8. Abbou, D., T. P. Vuong, R. Touhami, F. Ferrero, D. Hamzaoui, and M. C. E. Yagoub, "High-gain wideband partially reflecting surface antenna for 60 GHz systems," *IEEE Antennas and Wireless Propagation Letters*, vol. 16, pp. 2704–2707, 2017.

9. Xie, P., G. Wang, H. Li, and J. Liang, "A dual-polarized two-dimensional beam-steering Fabry–Pérot cavity antenna with a reconfigurable partially reflecting surface," *IEEE Antennas and Wireless Propagation Letters*, vol. 16, pp. 2370–2374, 2017.

10. Latif, S. I., L. Shafai, and C. Shafai, "Gain and efficiency enhancement of compact and miniaturised microstrip antennas using multi-layered laminated conductors," *IET Microwaves, Antennas & Propagation*, vol. 5, no. 4, pp. 402–411, March 21, 2011.

11. Attia, H., O. Siddiqui, L. Yousefi, and O. M. Ramahi, "Metamaterial for gain enhancement of printed antennas: Theory, measurements and optimization," *2011 Saudi International Electronics, Communications and Photonics Conference (SIECPC)*, Riyadh, pp. 1–6, 2011.

12. Li, D., Z. Szabo, X. Qing, E. P. Li, and Z. N. Chen, "A high gain antenna with an optimized metamaterial inspired superstrate," *IEEE Transactions on Antennas and Propagation*, vol. 60, no. 12, pp. 6018–6023, Dec. 2012.

13. Augustin, G., B. P. Chacko, and T. A. Denidni, "A zero-index metamaterial unit-cell for antenna gain enhancement," *2013 IEEE Antennas and Propagation Society International Symposium (APSURSI)*, Orlando, FL, pp. 126–127, 2013.

14. Kumar, S., L. Kurra, M. Abegaonkar, A. Basu, and Shiban K. Koul, "Multi-layer FSS for gain improvement of a wide-band stacked printed antenna," *2015 International Symposium on Antennas and Propagation (ISAP)*, Hobart, TAS, pp. 1–4, 2015.

15. Suthar, H., D. Sarkar, K. Saurav, and K. V. Srivastava, "Gain enhancement of microstrip patch antenna using near-zero index metamaterial (NZIM) lens," *2015 Twenty First National Conference on Communications (NCC)*, Mumbai, pp. 1–6, 2015.

16. Zhu, H., S. W. Cheung, and T. I. Yuk, "Enhancing antenna boresight gain using a small metasurface lens: Reduction in half-power beamwidth," *IEEE Antennas and Propagation Magazine*, vol. 58, no. 1, pp. 35–44, Feb. 2016.

17. Singh, Amit K., Mahesh P. Abegaonkar, and Shiban K. Koul, "High-gain and high-aperture-efficiency cavity resonator antenna using metamaterial superstrate," *IEEE Antennas and Wireless Propagation Letters*, vol. 16, pp. 2388–2391, June 2017.

18. Kurra, Lalithendra, Mahesh P. Abegaonkar, and Shiban K. Koul, "Equivalent circuit model of resonant EBG band stop filter," *IETE Journal of Research*, vol. 62, no. 1, pp. 17–26, Jan. 2016.
19. Singh, A. K., Mahesh P. Abegaonkar, and Shiban K. Koul, "Wide gain enhanced band high gain Fabry-Perot cavity antenna using ultrathin metamaterial surface," *Microwave and Optical Technology Letters*, pp. 1628–1633, Feb. 2019.
20. Qin, F. et al., "A triband low-profile high-gain planar antenna using Fabry–Perot cavity," *IEEE Transactions on Antennas and Propagation*, vol. 65, no. 5, pp. 2683–2688, May 2017.

5

High-Gain Antennas Using a Transmission-Type Metasurface

5.1 Introduction

High-gain compact antennas are one of the major requirements of modern wireless communication systems. High-gain antennas provide larger coverage with fewer chances of call drop and good quality of reception. This chapter reports the design and development of high-gain antennas using metamaterial lens. High-performance antennas have always attracted much attention due to different applications in EM engineering, especially the rapidly increasing demand of millimetre-wave communication systems. Metamaterials are a new kind of artificially engineered effective medium-based structure, which has been used to design a new type of antennas and enhance performance of traditional antennas. Using transformation optics [1], some high-directivity antennas based on anisotropic and inhomogeneous metamaterials have been proposed and analysed [2–4]. However, the high anisotropy or extreme parameters of the lens restrict the bandwidth and increase the loss of the antennas. Recently, 2D metamaterial MSs have been used to design new types of lens antennas for permanence improvement of conventional antennas. In [5], MS was used to collimate the radiation with minimal reflection loss. A cavity-excited Huygens' MS antenna has been proposed and demonstrated using the equivalence principle [6]. However, such MSs have relatively narrow bandwidth. Unique radiation effects may be obtained with CRLH structures due to their rich and unusual propagation properties. In addition, novel resonant-type metamaterial (MTM) antennas, based on the MTM resonators, are presented. The metamaterial is referred to as uniform when the elements in the periodic lattice are identical all over the plane [7–10]. In this chapter, we will discuss the design and development of uniform metamaterial lens to achieve enhanced directivity, enhanced gain and enhanced aperture efficiency. Several single-layer to multi-layer metamaterial lenses are designed and experimentally verified for their conversing nature and special properties. There are several types of metamaterial lenses reported by various researchers [11–20] for antenna characteristic enhancement. In addition, other approaches using

DOI: 10.1201/9781003045885-5

metamaterial lenses are also reported in the literature [21–23]. These lenses are self-sufficient in enhancing the radiation characteristics when placed at proper focal point of the lens.

5.2 Working Principle

The MS lens is designed using sub-wavelength transparent metamaterial UC. The highly transmissive MS is designed in a way to provide beam collimation on the MS top surface, resulting in maximum directive beam in the broadside direction. The total lens antenna directivity is a function of the microwave source feed antenna directivity and position of feed source.

The working principle of the MS lens is best understood by using the ray-tracing model shown in Figure 5.1.

A printed microwave source is used for easy integration with planar circuits. The waves are incident on MS lens at various incident angles (θ). The normal incident waves on MS lens pass through the lens without any deviation. The incident waves from oblique angles are converged by MS lens. The converging strength of lens depends on incidence angle and height of lens from microwave source. As the incidence angle increases, radiated power density on the side of the lens decreases. Since incidence angle is a function of location of the microwave source, the location of microwave source can also determine converging strength of the proposed MS lens.

The field distribution due to MS lens over radiator can be realized using the concept depicted in Figure 5.2. The equivalent circuit model of MS lens over radiator is shown in Figure 5.2(b). The parallel normalized inductive susceptance "$-jY_L$" in Figure 5.2(b) is due to MS lens. The input reflection coefficient "Γ_{in}" due to this configuration is given as

$$\Gamma_{in} = \frac{1 - Y_{in}}{1 + Y_{in}} = \frac{jY_1}{2 - jY_1}. \tag{5.1}$$

Here "Y_{in}" is the normalized input admittance given by $Y_{in} = 1 - jY_1$.

Consider electric field at $Z = 0$ (on the patch towards MS) with magnitude "E_o". When the EM wave propagates by "t" distance in Z direction towards MS, the E-field will be "$E_o\, e^{-ikt}$". After reflection by MS, E-field can be written as "$\Gamma_{in} E_o e^{-ikt}$". Due to multiple layers of MS as a superstrate over the radiator, multiple reflections will occur. Hence, E-field due to multiple reflections can be written as

$$E = E_o e^{-ikz} + \Gamma_{in} E_o e^{-ik(2t-Z)} + \Gamma_{in}^2 E_o e^{-ik(2t+Z)} + \Gamma_{in}^3 E_o e^{-ik(3t-Z)}. \tag{5.2}$$

Here "t" is the gap between MS layers, "$K = \dfrac{2\pi}{\lambda}$" is the wave number and "λ"

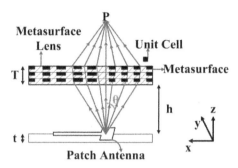

FIGURE 5.1
Ray-tracing model of small MS lens antenna.

Source: Copyright/used with permission of/courtesy of IET.

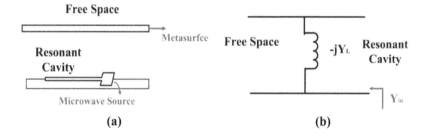

FIGURE 5.2
(a) MS lens over radiator schematic and (b) equivalent circuit diagram.

Source: Copyright/used with permission of/courtesy of IET.

is wavelength in the propagating medium. Solving the above equation, we get

$$E = \frac{E_o \left(e^{-ikz} + \Gamma_{in} e^{-ik(2t-Z)} \right)}{1 - \Gamma_{in}^2 e^{-ik(2t-Z)}}. \tag{5.3}$$

Putting "$t = 0$" and considering strong resonance condition, Equation 5.3 can be rewritten as

$$E = \frac{E_o \left(e^{-ikz} - e^{+ikz} \right)}{1 - e^{+ikz}} = E_o e^{-i\frac{kz}{2}} \left(\frac{\sin kz}{\sin \frac{kz}{2}} \right). \tag{5.4}$$

Equation 5.4 describes the E-field distribution on the top surface of MS lens. The strongest electric field is obtained at the centre of the MS lens and it deceases exponentially to the other sides. Similarly, H-field due to multiple-layer stacked MS superstrate can be analysed. Due to this, the radiation pattern on the MS lens surface is maximum at the centre of the MS and it decreases towards the sides. This will cause a convergence of the EM waves to the broadside direction by reducing the HPBW of the microwave source radiator. The MS lens antenna design is a complex problem and hence requires a significant and accurate analytical tool. CST microwave studio is used for all lens antenna analysis here. All the conditions mentioned earlier can be obtained only under strong resonance and matching conditions.

5.3 Design of an Ultrathin Miniaturized Metasurface for Wideband Gain Enhancement for C-Band Applications

In this section, a low-profile ultrathin miniaturized MS is designed and used to enhance radiation and transmission characteristics of wide-frequency C-band radiator. The lensing action of MS lens is extensively characterized in an anechoic chamber. A high-gain lens antenna is designed with wide enhanced gain bandwidth. The simulated and measured results are found to be in good agreement [24].

5.3.1 Metasurface and Wideband Enhanced-Gain Antenna Design

The planar low-profile ultrathin MS discussed in Section 2.5.3 is used here for designing a high-gain lens antenna. The modified double annular slot ring resonator (DASR) with metallic serrations (UC in Section 2.5.1) has measured wide pass-band characteristics from 4.25 GHz to 8.0 GHz with transmission bandwidth of 3.75 GHz. The size of MSUC is $0.166\lambda \times 0.166\lambda$, with planar MS thickness of 0.011λ making MS ultrathin and compact. The proposed new modified DASR UC MS is a wide pass-band MS and can be used to enhance radiation characteristics in wide frequency band. Placing MS over a microwave radiator to enhance radiation characteristics is reported in [20–27]. The MS should be placed at an optimum height to match focusing condition of microwave radiator. The optimum height "h" should be near half-wavelength at resonance frequency to achieve convergence of radiating fields, causing improvement in radiation and impedance characteristics [10–11]. A standard C-band WG port is considered as a wideband microwave radiator operating from 4.0 GHz to 8.0 GHz. The proposed MS is used to enhance radiation characteristics of proposed wideband microwave radiator. The MS is placed over a radiator at an optimum height to achieve maximum matching.

The antenna setup is shown in Figure 5.3. The optimum height to achieve wideband radiation convergence is obtained by varying "h" and measuring the transmission characteristics under normal incidence. The free space transmission characteristics of C-band horn WG port with and without MS at various heights "h" from radiating surface are plotted in

FIGURE 5.3
C-band WG port radiator with modified DASR metasurface.

Source: Copyright/used with permission of/courtesy of IEEE.

Figure 5.4. The transmission characteristic of C-band WG port without and with air gap using free space measurement transmission setup is shown in Figure 5.4. A free space path loss of about 25 dB is observed. The transmission characteristics of the C-band WG port is improved for wide frequency band when the height "h" is 25 mm, indicating improvement in radiation characteristics in the same band. Hence, the optimum height "h" considered is 25 mm. The optimum height is about $0.460\lambda_0$, where λ_0 is wavelength at resonant frequency 5.40 GHz. Hence, the focusing condition is satisfied.

5.3.2 Measurement of Radiation Characteristics

The radiation characteristics of C-band WG port without MS and C-band WG port with MS at optimum height of "$h = 25$ mm" are measured in an anechoic chamber using signal generator and spectrum analyser.

The measured radiation characteristics of both antenna setups for vertical linear electric field polarization at measured resonance frequency of 5.40 GHz are shown in Figure 5.5. A measured gain enhancement of 6.75 dB is observed at 5.4 GHz resonance frequency. The 3-dB HPBW of C-band WG port with MS is also improved by 8.5°.

The wideband radiation characteristics of C-band WG port without MS and with MS at optimum height "$h = 25$ mm" are measured for other frequencies from 4.0 GHz to 8.0 GHz in an anechoic chamber. The wide radiation characteristic is measured for "$h = 30$ mm" for other frequencies. The measured antenna gains of the C-band WG port without MS, the C-band WG port with MS at an optimum height of "$h = 25$ mm" and with "$h = 30$ mm" is compared in Figure 5.6. A wideband gain enhancement of 1.5 dB to 6.75 dB is achieved for optimum height "$h = 25$ mm" from 4.35 GHz to 7.45 GHz, indicating wideband enhanced gain bandwidth of 3.10 GHz with maximum

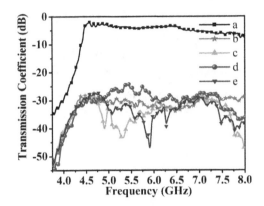

FIGURE 5.4
Measured transmission characteristic (a) C-band WG port without MS and without air gap; (b) C-band WG port in free space without MS in far-field region; (c) C-band WG port in free space with MS at "h = 20 mm" from the transmitter; (d) C-band WG port in free space with MS at "h = 25 mm" from the transmitter and (e) C-band WG port in free space with MS at "h = 30 mm" from the transmitter.

Source: Copyright/used with permission of/courtesy of IEEE.

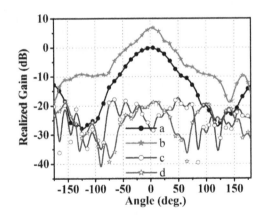

FIGURE 5.5
Measured radiation pattern in H-plane. (a) H.co, C-band WG port without MS; (b) H.co, C-band WG port with MS at optimum height "h = 25 mm", (c) H.cross, C-band WG port without MS and (d) H.cross, C-band WG port with MS at optimum height "h = 25 mm".

Source: Copyright/used with permission of/courtesy of IEEE.

enhanced gain of 6.75 dB at 5.4 GHz. Similar to the above, the new DASR UC MS at their optimum height "h" can be used with any wide frequency C-band microwave radiator, and similar type of wideband gain enhancement can be achieved.

A wideband gain enhancement of 1.5 dB to 6.75 dB with wide enhanced-gain-bandwidth of 3.15 GHz is successfully demonstrated. The designed

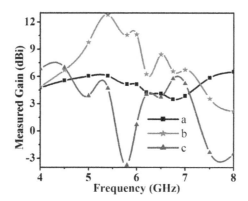

FIGURE 5.6
Measured Gain; (a) C-band WG port without MS; (b) C-band WG port with MS at "h = 25 mm" and (c) C-band WG port with MS at "h = 30 mm".

Source: Copyright/used with permission of/courtesy of IEEE.

MS can be used with any similar frequency band microwave radiator placed at an optimum height to enhance radiation and transmission characteristics. The designed antenna setup can be used for mobile microwave imaging and sensing, for high-gain satellite antennas and for 5G link antenna.

5.4 A Negative-Index Metamaterial Lens for Antenna Gain Enhancement

In this section, a transmission-type negative refractive index metamaterial surface acting as a planar surface lens is presented first. This designed MS is used to enhance the gain of an X-band patch antenna radiator [28]. The MS and high-gain lens antenna are characterized experimentally.

5.4.1 Design of the Metasurface and Working Principle

A low-profile transmission-type NIM MS with very high aperture efficiency is proposed. The proposed UC consists of two MCsRRs. Two concentric CRRs are designed first. To achieve compactness, the CRRs are loaded with capacitive slot in series. The capacitive slot length is more on inner CRR as compared to outer CRR. Slotted CRRs are called MCsRR. The inner MCsRR is more capacitive due to longer metallic slots, whereas outer MCsRR is more inductive due to smaller metallic slots. Due to normal incidence of the EM waves, the UC behaves as a resonant circuit. The resonant frequency due to MDCsRR UC is calculated using Equation 5.5.

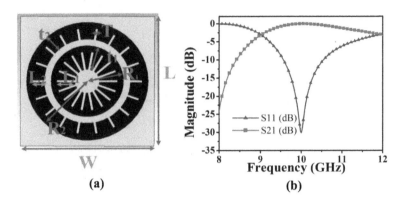

FIGURE 5.7
(a) Dimensions of proposed UC and (b) simulated transmission and reflection characteristic plot of the UC. The dimensions are $R_1 = 4.0$, $R_2 = 7.0$, $L_1 = 1.51$, $L_2 = 0.94$, $t_1 = 3.0$, $t_2 = 1.40$, $T = 0.20$, $L = 11.00$, $W = 11.00$ (all dimensions are in millimetres).

Source: Copyright/used with permission of/courtesy of IEEE.

$$f_0 = \frac{1}{2\pi\sqrt{L_{eq}C_{eq}}} \tag{5.5}$$

Here, "L_{eq}" is total equivalent inductance due to MDCsRR UC and "C_{eq}" is total equivalent capacitance due to MDCsRR UC. By changing metallic slot lengths "L_1 and L_2", resonant frequency can be changed. MDCSRR UC geometry with dimension is shown in Figure 5.7(a). The proposed UC is simulated under PEC and PMC boundary conditions as discussed in Section 2.4.1. The simulated transmission and reflection characteristics of the proposed MDCsRR UC are plotted in Figure 5.7(b). A simulated transmission band of 3 GHz from 8.9 GHz to 11.9 GHz is observed.

A transmission-type MS is designed using 5 × 5 array of MDCsRR as a UC. MS is fabricated on Neltec substrate with permittivity 3.2 and thickness of 0.762 mm. The fabricated MS is shown in Figure 5.8(a). Material property of MS is extracted using free space technique as reported in [29] and the same is shown in Figure 5.8(b). The proposed MS is NIM in the desired band of operation. When the propagating EM wave strikes the UC normally, the MDCsRR resonates at frequency "f_0".

The transmission and reflection characteristics of proposed MS are measured by placing fabricated MS in between two X-band WG ports placed in the far-field of each other as discussed in Chapter 2. Measured 3-dB transmission bandwidth of 2 GHz from 8.55 GHz to 10.55 GHz is observed from the measured response plot shown in Figure 5.9.

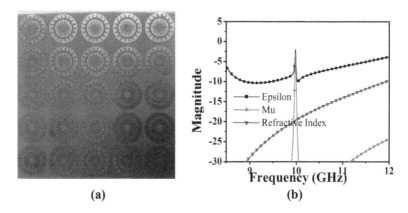

(a)   **(b)**

FIGURE 5.8
(a) The fabricated metasurface and (b) extracted material parameters of metasurface.

Source: Copyright/used with permission of/courtesy of IEEE.

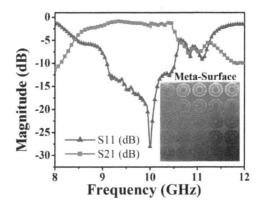

FIGURE 5.9
Measured transmission and reflection characteristic plot of metasurface.

Source: Copyright/used with permission of/courtesy of IEEE.

5.4.2 Design of the Metasurface Lens and Experimental Characterization

The simulated results obtained using CST microwave studio indicate that the conversing strength of single-layer proposed MS is small. To increase the conversing strength of MS, two similar proposed MSs are stacked together, and a hybrid lens is formed. This double-layer stacked MS is found to have higher conversing strength than that of the single-layer MS. To act as a lens, the double-layer stacked MS should be placed over MPA acting as microwave

FIGURE 5.10
Fabricated prototype of small NIM lens antenna.

Source: Copyright/used with permission of/courtesy of IEEE.

source at optimum height where maximum matching between source and MS can be achieved.

An MPA operating at 10.06 GHz is fabricated on the same substrate to act as a microwave source. Next, a high-gain antenna is designed by placing double stacked (DS) layer of proposed NIM MS over MPA. To find out the optimum height of NIM MS over MPA, parametric variation is carried out, and the optimum height in the proposed design is found to be 16.50 mm, which is $0.551\lambda_0$ (about half of the resonating wavelength). Spacers are used to provide optimum height of NIM lens. The final designed NIM lens antenna is shown in Figure 5.10, where "$h = 16.50$ mm".

The reflection coefficient of patch antenna and NIM lens antenna prototype is measured using VNA. The measured and simulated return loss of MPA and NIM antenna is compared in Figure 5.11. A good agreement between simulated and measured results is obtained.

The radiation characteristics of NIM lens antenna and MPA are measured in an anechoic chamber. The measured and simulated E-plane and H-plane radiation characteristics are plotted in Figure 5.12(a) and Figure 5.12(b), respectively. Measured gain enhancement of 8.55 dB in H-plane and 6.20 dB in E-plane is achieved whereas measured HPBW of NIM antenna is reduced by 25° and 42.5°, as compared to HPBW of MPA in H-plane and E-plane, respectively. The reduction in HPBW indicates convergence of radiation pattern, resulting in gain enhancement in broadside direction. When the designed DS MS is placed over a microwave source at an optimum height having broad radiation pattern, the MS starts acting like a planar surface lens and causes convergence of radiation pattern in the broadside direction. This results in reduction in HPBW and causes gain enhancement. The 3D simulated radiation pattern of MPA and NIM lens antenna is shown in Figure 5.13.

A transmission-type NIM MS acting as a conversing lens is successfully demonstrated. DSNIM MS over MPA causes gain enhancement of MPA by 8.55 dB in H-plane and 6.20 dB in E-plane. A high-gain antenna is designed using DS NIM MS over MPA having a calculated antenna gain of 13.35 dB. This proposed high-gain antenna can be used for microwave imaging and line-of-sight communication.

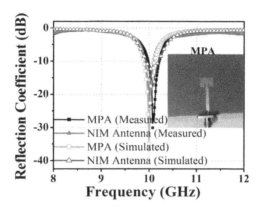

FIGURE 5.11
Measured and simulated reflection coefficients of MPA and NIM lens antenna.

Source: Copyright/used with permission of/courtesy of IEEE.

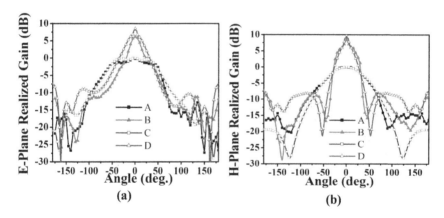

FIGURE 5.12

Measured and simulated radiation pattern of MPA and NIM antennas (a) E-plane (b) H-plane, where A is MPA measured, B is NIM antenna measured, C is MPA simulated and D is NIM antenna simulated.

Source: Copyright/used with permission of/courtesy of IEEE.

5.5 Design of a Compact Near-Zero Index Metasurface Lens with High Aperture Efficiency for Antenna Radiation Characteristic Enhancement

In this section, a high-gain, aperture-efficient, low-profile compact, near-zero index (NZI) MS lens antenna with enhanced radiation characteristic is

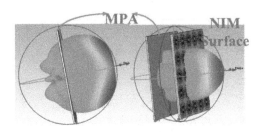

FIGURE 5.13
Simulated 3D radiation pattern of MPA and NIM antenna.

Source: Copyright/used with permission of/courtesy of IEEE.

proposed. The focusing mechanism of single-layer to multiple-layer stacked MS is studied. A compact high-gain MS lens antenna with high aperture efficiency of around 97 per cent for radiation characteristic enhancement of MPA radiator is designed [30]. The focusing mechanism of single-layer MS to multiple-layer stacked MS is experimentally demonstrated.

5.5.1 Design of the Metasurface

The proposed metamaterial UC is a compact highly inductive and capacitive modified double circular ring resonator (MDCRR). A double-circular ring resonator (DCRR) with pass-band centre frequency at 13.75 GHz is designed first. The DCRR is modified with additional capacitive slots to achieve compactness having pass-band centre frequency at 9.97 GHz. The additional capacitive slot causes compactness of 27.50 per cent. The proposed MDCRR UC is complementary to double annular slot ring resonator (DASR) presented in Chapter 2, which is a reflection-type MS. The DASR MS works on cavity resonance principle as demonstrated in Chapter 4. The proposed MDCRR MS works on lensing effect of MS. The proposed UC is designed on Rogers RT/duroid 5880 substrate having permittivity 2.2 and thickness of 0.762 mm with dimension of $a{\times}b$. The detailed geometry of UC with dimensional parameters is given in Figure 5.14(a). The proposed UC is simulated with PEC and PMC boundary conditions excited by a plane wave using WG port as shown in Figure 5.14(b).

Simulated transmission and reflection parameters with phase plot are given in Figure 5.15(a), (b), respectively. The proposed metamaterial UC has simulated 3-dB transmission band from 8.47 GHz to 13.00 GHz with resonance frequency at 9.97 GHz.

The simulated phase characteristic plot after port de-embedding is shown in Figure 5.15(b), having reflection phase of $-120°$ and a transmission phase of $-15.11°$ at centre frequency of 9.97 GHz. EM material property of proposed UC is extracted using algorithm reported in [11] and is shown in Figure 5.15(c). The refractive index of the proposed MS is found to be near zero. Hence, the proposed MS is NZI metamaterial (NZIM).

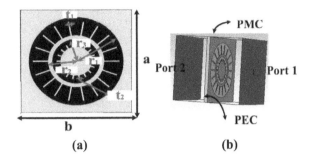

FIGURE 5.14

(a) MDCRR UC geometry with dimensional parameters and (b) simulation boundary conditions. The dimensions are $a = 13.00$, $b = 13.00$, $r_1 = 4.50$, $r_2 = 3.00$, $r_3 = 2.50$ and $t_1 = 0.22 = t_2$ (all dimensions are in millimetres).

Source: Copyright/used with permission of/courtesy of IET.

A high-performance transmissive MS is designed using 4 × 4 array of proposed completely transparent MDCRR UC. The proposed MS prototype is fabricated on Rogers RT/duroid 5880 substrate having permittivity 2.2 and thickness of 0.762 mm. The designed MS with dimensional parameters is given in Figure 5.16(a), where $L = W = 52$ mm. The fabricated prototype MS is shown in Figure 5.16(b). The transmission and reflection characteristics of the proposed MS are measured using free space technique by placing MS between two standard X-band WG port radiators as discussed in Chapter 2. The simulated and measured transmission and reflection characteristics of the proposed MS are shown in Figure 5.16(c). As observed, the measured results are in good agreement with simulated results. The fabricated MS has measured 3-dB transmission band of 3.10 GHz from 8.80 GHz to 11.90 GHz. The return loss characteristics show small variation with respect to simulated one with maximum dip at 10.15 GHz. The variation mentioned earlier is due to misalignment and fabrication inaccuracies.

5.5.2 Design of a Metasurface Lens and Characterization

In this section, a small NZI MS lens is designed using proposed MS to achieve high gain and high aperture efficiency by converging the signal in the direction of boresight. Lensing effect of single-surface and multiple-surface stacked MS is studied first, and finally three similar layers of proposed MS are stacked together to form highly directive MS lens. The lens action of the proposed MS lens is explained by using the ray-tracing model as shown in Figure 5.1. The lensing effect of proposed MS lens is studied by placing single layer, double-layer stacked and triple-layer stacked of proposed MS over MPA. The return loss characteristic and radiation pattern at different optimum heights "h" are studied.

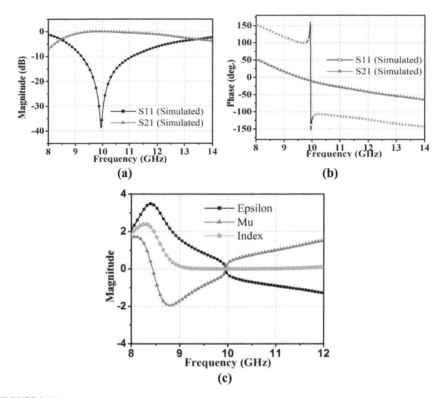

FIGURE 5.15

(a) Simulated transmission and reflection coefficient magnitude; (b) simulated transmission and reflection coefficient phase plot and (c) extracted material parameters epsilon, mu and refractive index.

Source: Copyright/used with permission of/courtesy of IET.

Lensing effect of the proposed single-layer MS placed over an MPA operating at 10 GHz acting as a source is studied first. The optimum height of MS is selected by doing parametric variation of "*h*" to achieve maximum matching. The reflection coefficient of single-layer MS over MPA is measured at different heights. Simulated and measured reflection coefficients for different height variations of MS are plotted in Figure 5.17(a). The small mismatch in measured and simulated return loss is due to loading effect.

The radiation patterns of MPA and MPA with single-layer MS are measured in an anechoic chamber. The measured radiation pattern of patch antenna and patch antenna with single MS in H- and E-planes at different heights "*h*" is plotted in Figure 5.17 (b), (c), respectively. The optimum height obtained is 15 mm, where maximum matching with maximum enhanced gain is achieved. A measured gain enhancement of 0.17 dB in H-plane and 0.15 dB in E-plane is obtained at optimum height "*h* = 15 mm".

The lensing effect of double-layer stacked MS placed over MPA is studied. Optimum height "*h*" is obtained by simulated and measured parametric

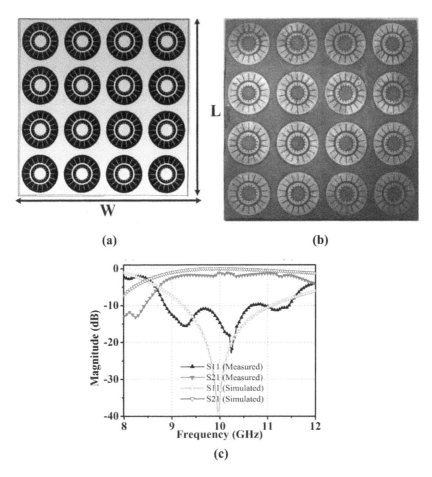

FIGURE 5.16

(a) Proposed metasurface design with dimensions; (b) fabricated prototype MDCRR MS and (c) measured and simulated transmission and reflection characteristics ($L = W = 52$ mm).

Source: Copyright/used with permission of/courtesy of IET.

variations as shown in Figure 5.18(a). The mismatch in measured and simulated return loss is due to loading effect. The measured radiation pattern of MPA and double-layer stacked MS over MPA at different heights "h" in H- and E-planes is plotted in Figure 5.18 (b), (c), respectively. A measured gain enhancement of 7.57 dB in H-plane and 7.3 dB in E-plane is obtained at an optimum height of 19.75 mm. The E-field and H-field is found to be converging towards the broadside.

Finally, lensing effect of triple-layer stacked MS over MPA is studied. The optimum height "$h = 16.50$ mm" is obtained through simulations and measured parametric variation of "h" as shown in Figure 5.19(a). Loading effect of MS over MPA causes small mismatch in measured and simulated reflection coefficients.

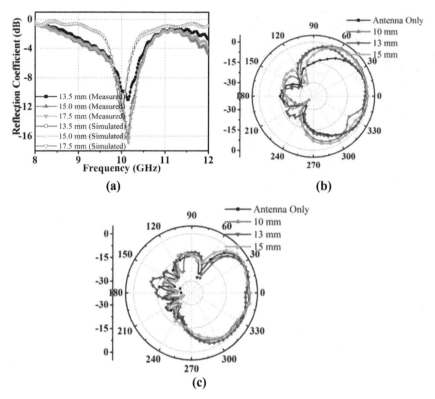

FIGURE 5.17

Single-layer MS over MPA with height variation. (a) Measured and simulated reflection coefficients with height variation; (b) measured H-plane radiation pattern and (c) measured E-plane
radiation pattern.

Source: Copyright/used with permission of/courtesy of IET.

The measured radiation patterns of MPA and triple-layer stacked MS lens
antenna at different heights "*h*" in H-plane and E-plane are plotted in
Figure 5.19(b), (c), respectively. Measured gain enhancement of 10.70 dB in
H-plane and 11.67 dB in E-plane is observed. The E- and H-plane radiation
patterns are found to be further converging towards the broadside.

To understand the phenomenon discussed earlier, simulated H-field distribution of patch antenna, patch antenna with single-layer MS superstrate,
double-layer stacked MS superstrate and triple-layer stacked MS superstrate
is shown in Figure 5.20. Simulated results show that rate of convergence of
the H-field is maximum due to triple-layer stacked MS over MPA.

To analyse further the effect of more number of stacked MS layers over MPA,
analytical study is done and a graph of simulated antenna gain of MPA with
stacked MS versus number of stacked MS is plotted in Figure 5.21. Maximum
gain of stacked MS over MPA is achieved when number of stacked layers is

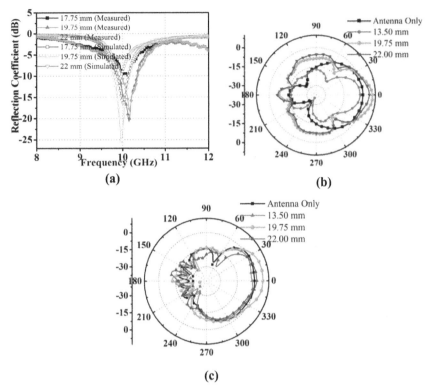

FIGURE 5.18

Double-layer MS over MPA with height variation. (a) Measured and simulated reflection coefficients with height variation; (b) measured H-plane radiation pattern and (c) measured E-plane radiation pattern.

Source: Copyright/used with permission of/courtesy of IET.

three. As the number of stacked layers over MPA increases, simulated gain of MS over MPA decreases. Hence, the optimum number of MS for stacking to design a highly converging microwave lens is three here. The E- and H-plane measured radiation patterns, simulated H-field distribution and results, discussed earlier and presented in Figure 5.21, suggest that the highest convergence of E- and H-fields is obtained by triple-layer stacked MS over MPA. To study the effect of dielectric thickness of MS lens, MS of thickness three times that of single-layer MS with same MDCRR UC on the top is designed and placed at an optimum height of 15.50 mm over MPA. Simulated gain of 8.75 dB, indicating gain enhancement of 3.25 dB, is observed. It can be concluded from this analysis that the actual gain enhancement is due to triple-layer stacked MS lens.

5.5.3 Design of a High-Gain-Single-Surface Lens Antenna

A microwave source is designed by fabricating an MPA operating at 10 GHz on Rogers RT/Duroid 5880 substrate having permittivity 2.2 and thickness of

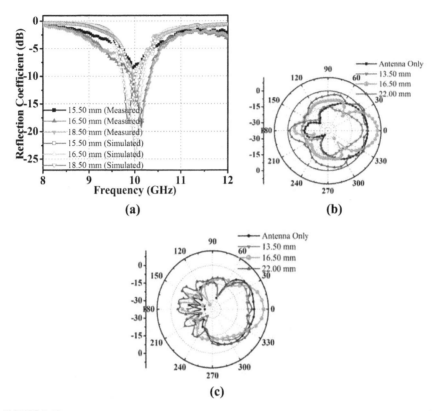

FIGURE 5.19

Triple-layer MS over MPA with height variation; (a) measured and simulated reflection coefficients with height variation; (b) measured H-plane radiation pattern and (c) measured E-plane radiation pattern.

Source: Copyright/used with permission of/courtesy of IET.

0.762 mm. A prototype NZI MS lens antenna is fabricated by placing single-surface triple-layer stacked NZI MS lens over MPA at analysed optimum height of 16.50 mm as shown in Figure 5.22(a). The return loss is measured using Agilent PNA model E8364 VNA. The simulated and measured return loss of MPA and MS lens antenna is compared in Figure 5.22(b). The measured and simulated results are in good agreement. The measured return loss of MS lens antenna is slightly different from simulated return loss because MS lens antenna is very sensitive to optimum height "h" and due to the loading effect of MS.

The radiation characteristics of MPA and NZI MS lens antenna for vertical linear electric field polarization and magnetic field polarization are measured in an anechoic chamber using spectrum analyser and signal generator. Simulated and measured gain patterns for vertical linear electric field polarization of MPA and NZI small MS lens antenna are plotted in Figure 5.23(a). The H-plane realized gain pattern is highly converging towards broadside,

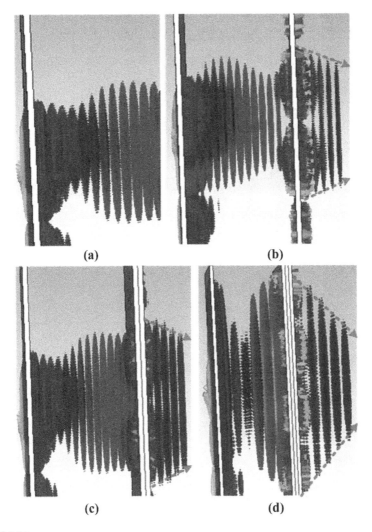

FIGURE 5.20
Simulated H-field distribution; (a) patch antenna; (b) patch antenna with single-layer MS; (c) patch antenna with double-layer MS and (d) patch antenna with triple-layer MS.

Source: Copyright/used with permission of/courtesy of IET.

causing gain enhancement of 10.70 dB at 10.15 GHz frequency. The E-plane gain pattern of MPA and NZI MS lens antenna is plotted in Figure 5.23(b). The simulated and measured patterns are in good agreement. The gain pattern is converging towards broadside, causing gain enhancement of 11.67 dB at operating frequency 10.15 GHz. The measured gain of NZI small MS lens antenna is 15.49 dB. Measured 3-dB gain bandwidth of 1.5 GHz is obtained. A total phase of −720.50° (due to 2h phase of −402.0°, due to 2t phase of −18.5°, due to metal ground −180° and due to MS −120°), close to 4π is generated at

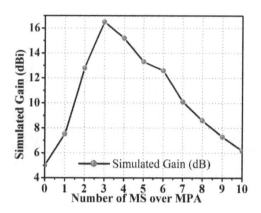

FIGURE 5.21

MS lens antenna simulated gain with number of stacked MS over MPA.

Source: Copyright/used with permission of/courtesy of IET.

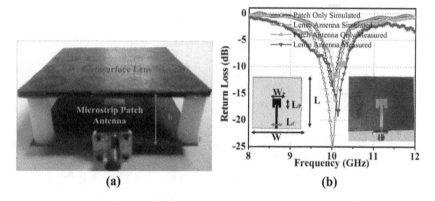

FIGURE 5.22

Fabricated prototype small NZI MS lens antenna. (a) Fabricated small NZI MS lens antenna and (b) simulated and measured reflection coefficient of patch antenna and small NZI MS lens antenna.

Source: Copyright/used with permission of/courtesy of IET.

MS causing coherent beam generation resulting into gain enhancement [31]. To summarize the focusing mechanism of the proposed MS lens antenna, simulated and measured gain of antennas with simulated and measured HPBWs are compared in Table 5.1. Applying MS over MPA causes reduction in HPBW, resulting in gain enhancement in boresight direction.

The broadside simulated gain of NZI small MS lens antenna due to triple-layer stacked MS lens and various orders of misalignment of stacked MS are plotted in Figure 5.24. It is observed that a wide 3-dB gain bandwidth of 1.10 GHz with maximum simulated gain of 16.40 GHz can be obtained due to perfectly aligned triple-layer stacked MS lens. The simulated gain due to misalignment of top MS in steps of 2.5 mm along negative X-axis direction,

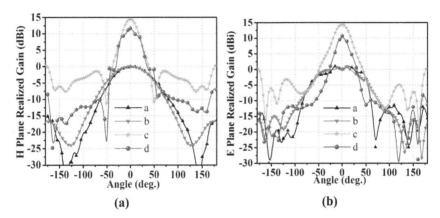

FIGURE 5.23
Simulated and measured radiation patterns (a) H-plane and (b) E-plane; a – patch antenna measured; b – patch antenna simulated; c – small metasurface lens antenna simulated; d – small metasurface lens antenna measured.

Source: Copyright/used with permission of/courtesy of IET.

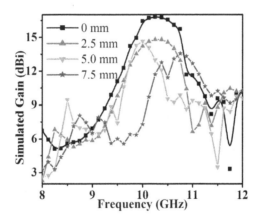

FIGURE 5.24
Simulated gain due to top MS layer variation along negative X-axis in step of 2.5 mm.

Source: Copyright/used with permission of/courtesy of IET.

keeping the bottom and middle surfaces fixed, is plotted in Figure 5.24. Similar results are obtained due to middle and bottom layer variation, keeping other two layers fixed. It is observed that misalignment causes significant reduction in broadside gain due to improper excitation of the UC.

To characterize the performance of the proposed antenna, the maximum broadside gain of effective radiating aperture is calculated as

$$G_{max} = (\frac{4\pi}{\lambda_O{}^2})A_{eff}E_f \tag{5.6}$$

TABLE 5.1

Simulated and Measured Gain with HPBW in an E-Plane

Antenna	Simulated Gain (dB)	Measured Gain (dB)	Simulated HPBW	Measured HPBW
MPA	5	3.85	82.9°	78°
MPA with single-layer MS	7.5	4.02	79.0°	76.5°
MPA with double-layer MS	12.7	10.20	34.5°	50°
MPA with triple-layer MS	16.4	15.49	30.2°	40°

TABLE 5.2

Gain Comparison with G_{max}

Antenna	Optimum Height (mm)	Gain Enhancement (dB)	% of G_{max}
MPA			24.06
MPA with single-layer MS	15.00	0.17	25.12
MPA with double-layer MS	19.75	6.35	63.75
MPA with triple-layer MS	16.50	11.64	96.80

In Equation 5.6, A_{eff} is effective aperture size of MS superstrate and E_f is radiation efficiency. The measured gain is compared with maximum gain corresponding to effective radiating aperture (G_{max}) size of 52 mm × 52 mm in Table 5.2. The simulated HPBW of NZI small MS lens over MPA for E- and H-planes is 30.2° and 28.8°, respectively. The measured HPBW of the same antenna for E- and H-planes is 40° and 25°, respectively. Applying triple-layer NZI small MS lens over MPA causes focusing of radiation energy by decreasing HPBW. The H-plane is more focused as compared to E-plane. Measured FBR of proposed small NZI MS lens antenna is improved by 3.2 dB in H-plane and by 8 dB in E-plane as observed from Figure 5.23(a), (b), respectively.

An NZI MS lens is successfully demonstrated and lensing effect is verified experimentally. Next, a low-profile high-gain antenna with enhanced gain and improved FBR using proposed NZI MS lens is successfully demonstrated. It is observed that the proposed MS lens converges E- and H-fields, causing gain enhancement by 10.70 dB and 11.67 dB in H- and E-planes, respectively. The calculated gain of the proposed small MS lens antenna is 15.49 dB, achieving 97.50 per cent of G_{max}, indicating that the proposed NZI MS lens antenna is highly aperture-efficient. The FBR is improved by 8 dB and 3 dB in E- and H-planes, respectively.

The proposed small MS lens antenna is compared with other similar reported lens antennas in Table 5.3 and is found to have more gain enhancement, more aperture efficiency and is very compact.

TABLE 5.3

Comparison of Proposed NZI Small MS Lens Antenna with Other Similar Reported MS Antennas

Ref.	Height of MS over MPA (λ_o)	Operating Frequency (GHz)	Dimension $L \times W$ (mm²)	Gain Enhancement (dB)	Antenna Gain (dBi)	Calculated Aperture Efficiency in Terms of G_{max}
[24]	0.873	10.0	180 × 180	7.8	11.43	67.87
[25]	0.488	12.4	60 × 60	7.2	13.75	69.08
[26]		9.5	144 × 144	8.0	21.75	50.00
[27]	0.097 (SIW)	9.73	89 × 50	5.8	8.5	49.29
[28]	1	10	104 × 104	9.2	15.6	73.13
Proposed Work	0.558	10.15	52× 52	11.7	15.49	97.50

5.6 Conclusion

In this chapter, microwave MS lens for C-band and X-band applications is proposed. The designed MS lens is low-profile high aperture efficiency lens, causing gain enhancement of 1.5 dB to maximum 11.67 dB in the broadside direction. A maximum gain enhancement of 11.67 dB, achieving 97.50 per cent of G_{max} due to effective radiating aperture is achieved, indicating the designed MS lens is highly aperture-efficient. High-gain antennas are designed by placing these MS lenses over microwave radiator at an optimum height.

In Chapter 6, linear and radial GIMS lenses for beam steering of microwave radiator with gain enhancement are presented.

References

1. Pendry, J. B., D. Schurig, and D. R. Smith, "Controlling electromagnetic fields," *Science*, vol. 312, pp. 1780–1782, June 2006.

2. Wu, Q. et al., "Transformation optics inspired multibeam lens antennas for broadband directive radiation," *IEEE Transactions on Antennas and Propagation*, vol. 61, no. 12, pp. 5910–5922, Dec. 2013.

3. Jiang, W. X., T. J. Cui, H. F. Ma, X. M. Yang, and Q. Cheng, "Layered high-gain lens antennas via discrete optical transformation," *Applied Physics Letters*, vol. 83, no. 22, p. 221906, 2008.

4. Aghanejad, I., H. Abiri, and A. Yahaghi, "Design of high-gain lens antenna by gradient-index metamaterials using transformation optics," *IEEE Transactions on Antennas and Propagation*, vol. 60, no. 9, pp. 4074–4081, Sep. 2012.
5. Pfeiffer, C. and A. Grbic, "Planar lens antennas of subwavelength thickness: Collimating leaky-waves with metasurfaces," *IEEE Transactions on Antennas and Propagation*, vol. 63, no. 7, pp. 3248–3253, July 2015.
6. Epstein, A., J. P. S. Wong, and G. V. Eleftheriades, "Cavity-excited Huygens' metasurface antennas for near-unity aperture illumination efficiency from arbitrarily large apertures," *Nature Communications*, vol. 7, Jan. 2016, no. 10360.
7. Minatti, G., F. Caminita, M. Casaletti, and S. Maci, "Spiral leaky-wave antennas based on modulated surface impedance," *IEEE Transactions on Antennas and Propagation*, vol. 59, no. 12, pp. 4436–4444, 2011.
8. Patel, A. M. and A. Grbic, "A printed leaky-wave antenna based on a sinusoidally-modulated reactance surface," *IEEE Transactions on Antennas and Propagation*, vol. 59, no. 6, pp. 2087–2096, 2011.
9. Minatti, G., M. Faenzi, E. Martini, F. Caminita, P. De Vita, D. Gonzalez-Ovejero, M. Sabbadini, and S. Maci, "Modulated metasurface antennas for space: Synthesis, analysis and realizations," *IEEE Transactions on Antennas and Propagation*, vol. 63, no. 4, pp. 1288–1300, 2015.
10. Fong, B. H., J. S. Colburn, J. J. Ottusch, J. L. Visher, and D. F. Sievenpiper, "Scalar and tensor holographic artificial impedance surfaces," *IEEE Transactions on Antennas and Propagation*, vol. 58, no. 10, pp. 3212–3221, 2010.
11. Minatti, G., S. Maci, P. De Vita, A. Freni, and M. Sabbadini, "A circularly-polarized Isolux antenna based on anisotropic metasurface," *IEEE Transactions on Antennas and Propagation*, vol. 60, no. 11, pp. 4998–5009, 2012.
12. Faenzi, M., F. Caminita, E. Martini, P. De Vita, G. Minatti, M. Sabbadini, and S. Maci, "Realization and measurement of broadside beam modulated metasurface antennas," *IEEE Antennas and Wireless Propagation Letters*, vol. 15, pp. 610–613, 2016.
13. Pandi, S., C. A. Balanis, and C. R. Birtcher, "Design of scalar impedance holographic metasurfaces for antenna beam formation with desired polarization," *IEEE Transactions on Antennas and Propagation*, vol. 63, no. 7, pp. 3016–3024, 2015.
14. Casaletti, M., M. Śmierzchalski, M. Ettorre, R. Sauleau, and N. Capet, "Polarized beams using scalar metasurfaces," *IEEE Transactions on Antennas and Propagation*, vol. 64, no. 8, pp. 3391–3400, 2016.
15. Sabbadini, M., G. Minatti, S. Maci, and P. De Vita, "Method for designing a modulated metasurface antenna structure," *Patent WO 2015090351 A1*, June 25, 2015.
16. Maci, S., G. Minatti, M. Casaletti, and M. Bosiljevac, "Metasurfing: Addressing waves on impenetrable metasurfaces," *IEEE Antennas and Wireless Propagation Letters*, vol. 10, pp. 1499–1502, 2011.
17. Martini, E. and S. Maci, "Metasurface transformation theory," in *Transformation Electromagnetics and Metamaterials*, ed by D. H. Werner, D. H. Know, Springer, London, pp. 83–116, 2013.
18. Pfeiffer, C. and A. Grbic, "A printed, broadband Luneburg lens antenna," *IEEE Transactions on Antennas and Propagation*, vol. 58, no. 9, pp. 3055–3059, 2010.
19. Bosiljevac, M., M. Casaletti, F. Caminita, Z. Sipus, and S. Maci, "Non-uniform metasurface Luneburg lens antenna design," *IEEE Transactions on Antennas and Propagation*, vol. 60, no. 9, pp. 4065–4073, 2012.

20. Pfeiffer, C. and A. Grbic, "Metamaterial Huygens' surfaces: Tailoring wave fronts with reflectionless sheets," *Physical Review Letters*, vol. 110, no. 19, p. 197401, 2013.

21. Selvanayagam, M. and G. Eleftheriades, "Discontinuous electromagnetic fields using orthogonal electric and magnetic currents for wavefront manipulation," *Optics Express*, vol. 21, no. 12, pp. 14409–14429, 2013.

22. Yu, N., P. Genevet, F. Aieta, M. A. Kats, R. Blanchard, G. Aoust, J.-P. Tetienne, Z. Gaburro, and F. Capasso, "Flat optics: Controlling wavefronts with optical antenna metasurfaces," *IEEE Journal of Selected Topics in Quantum Electronics*, vol. 19, no. 3, pp. 4700423–4700423, 2013.

23. Kildal, P. S., E. Alfonso, A. Valero-Nogueira, and E. Rajo-Iglesias, "Local metamaterial-based waveguides in gaps between parallel metal plates," *IEEE Antennas and Wireless Propagation Letters*, vol. 8, pp. 84–87, 2009.

24. Li, D., Z. Szabo, X. Qing, E. P. Li, and Z. N. Chen, "A high gain antenna with an optimized metamaterial inspired superstrate," *IEEE Transactions on Antennas and Propagation*, vol. 60, no. 12, pp. 6018–6023, Dec. 2012.

25. Yuan, H. Y., S. B. Qu, J. Q. Zhang, J. F. Wang, H. Y. Chen, H. Zhou, Z. Xu, and A. X. Zhang, "A metamaterial-inspired wideband high-gain FABRY–Perot resonator microstrip patch antenna," *Microwave and Optical Technology Letters*, vol. 58, no. 7, pp. 1675–1678, 2016.

26. Zhu, H., S. W. Cheung, and T. I. Yuk, "Enhancing antenna boresight gain using a small metasurface lens: Reduction in half-power beamwidth," *IEEE Antennas and Propagation Magazine*, vol. 58, no. 1, pp. 35–44, Feb. 2016.

27. Pandit, S., A. Mohan, and P. Ray, "A low-profile high-gain substrate-integrated waveguide-slot antenna with suppressed cross polarization using metamaterial," *IEEE Antennas and Wireless Propagation Letters*, vol. 16, pp. 1614–1617, 2017.

28. Li, H., G. Wang, H. X. Xu, T. Cai, and J. Liang, "X-band phase-gradient metasurface for high-gain lens antenna application," *IEEE Transactions on Antennas and Propagation*, vol. 63, no. 11, pp. 5144–5149, Nov. 2015.

29. Singh, Amit K., Mahesh P. Abegaonkar, and Shiban K. Koul, "Ultrathin miniaturized meta-surface for wide band gain enhancement," *2017 IEEE Asia Pacific Microwave Conference (APMC)*, Kuala Lumpar, pp. 383–386, 2017.

30. Singh, Amit K., Mahesh P. Abegaonkar, and Shiban K. Koul, "A negative index metamaterial lens for antenna gain enhancement," *2017 International Symposium on Antennas and Propagation (ISAP)*, Phuket, pp. 1–2, 2017.

31. Singh, Amit K., Mahesh P. Abegaonkar, and Shiban K. Koul, "A compact near zero index metasurface lens with high aperture efficiency for antenna radiation characteristic enhancement," *IET Microwaves, Antennas and Propagation*, vol. 13, no. 8, pp. 1248–1254, March 2019.

6

Beam Steerable High-Gain Antennas Using a Graded Index Metamaterial Surface

6.1 Introduction

Beam steerable high-gain antennas are important to achieve wide area coverage with less probability of error. To achieve these desired characteristics by a communication network, wide beam scanning with high gain is required. Beam steering antennas are an attractive solution for automotive radar systems, point-to-point and point-to-multipoint communication systems. Phased array aperture type antennas can provide adaptive beam scanning and beam shaping. These high-gain antennas can be used at base stations. However, the drawbacks of these systems are complexity, pencil beam shape and losses in the beam forming networks. The aperture antennas, having spatial localized phase shifters, can generate steered EM response with shaped beam, sharp cut-off and low loss. The local phase shifters with variable phase characteristics can be designed by modulating the characteristics of sub-wavelength metamaterial UCs. The largest dimension of the antenna can be of several wavelengths, up to 15λ. The local period of the global modulation is normally of the order of one wavelength. The lattice of the MS lens consists of pixel elements (UC) with a size of approximately $\lambda/10$. The specific features added to the patches to control polarization can be of the order of $\lambda/50$, which is approximately the size of the discretization triangles required in a method of moment mesh. Hence, to do the full analysis of the structure, complete full-wave analysis or solver will be required.

A phase-gradient MS lens is designed in [1] using modulated circular disc UC placed on the top of radiator to focus the radiated beam, resulting in gain enhancement with reduced beam width. Beam tilting of radiator is achieved using phase-modulated MS and negative indexed MS loading in [2–5]. Beam control of a substrate-integrated slot array antenna using MS to achieve wide beam scanning with gain enhancement is presented in [6]. The Luneburg lens, using high-impedance metamaterial surface and fused filament additive manufacturing process, is designed in [7, 8] to achieve beam focusing and beam steering. A bifocal Fresnel lens is designed using polarization-sensitive

DOI: 10.1201/9781003045885-6

MS in [9]. Beam steering of MPA radiator with gain enhancement is achieved using mechanical movement of hybrid graded-phase MS and passive turning GIMS in [10–11]. The design and realization of flat top-shaped beam antenna is proposed using optimization of amplitude and phase of array elements, modifying amplitude and phase along feed of array element and near-field phase transformation of double-layer antenna in [12–14].

6.2 Working Principle

The MS is designed using periodic arrangement of fundamental sub-wavelength UCs. The UC acts like a local atom for designing the complete material. The designed metamaterial may have transmissive or reflective characteristics depending on UC characteristics. The periodic arrangement of UC in a plane gives a uniform MS and non-periodic arrangement of UC gives a nonuniform MS. The transmission and reflection magnitude and phase response of a UC depend on the geometry and dimensions of the UC. Continuously changing geometry causes continuous change in transmission magnitude and phase of the UC, if the UC is of transmission type. The continuous change in transmission magnitude and phase causes continuous change in refractive index. As continuous changing dimension UC geometry is placed along a line in a plane, continuous change in refractive index occurs. As the change in dimension of the UC is constant and graded along a line in a plane, MS is called linear GIMS LGIMS. Similarly, if the graded changing dimension UC is placed along an angular line or radius of a circular plane, it is called angular/radial GIMS (AGIMS/RGIMS). The graded refractive index material can cause graded phase change in the radiated beam of a primary microwave radiator.

The working principle of a GIMS lens is explained using Figure 6.1. The proposed lens antenna configuration consists of a single-layer planar linear graded index modulated MS placed over a microwave radiator at an optimum height, generating modulated beam direction. The radiated EM wave by the radiator is converged by the GIMS lens in the direction of angle "θ". The radiated beam direction "θ" depends on the fundamental scatterer (UC) phase distribution profile. To control the broadside beam direction, the phase *profile of the UC can be controlled*. The modulated magnitude and phase profile of the MS placed over radiator can generate modulated direction of the radiated main beam. The beam tilt angle "θ" is proportional to linear phase gradient $\dfrac{d\Phi}{dx}$ generated by GIMS, where "Φ" is transmission phase of the UC and "x" is the axis along which change is graded as shown in Figure 6.1. A single-layer LGIMS having linear phase gradient can steer the main

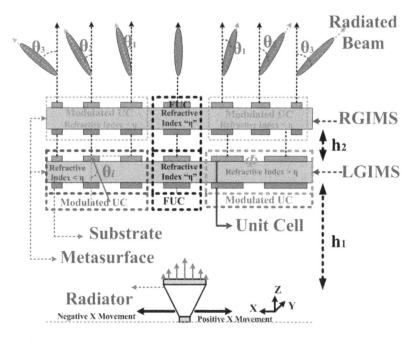

FIGURE 6.1

Working principle of GIMS lens antenna to achieve beam steering.

Source: Copyright/used with permission of/courtesy of IEEE.

beam of radiator along a conical plane with a vertex angle of "2θ", by moving about the radiator's parallel axis, changing the angle of incidence "θ_i". The beam steering angle "θ" depends on wave incidence angle "θ_i" and phase gradient $\dfrac{\mathrm{d}\Phi}{\mathrm{d}x}$. The steering angle "$\theta$" is given by following equation:

$$\sin\theta = \sin\theta_i + \frac{\lambda_o}{2\pi}\left(\frac{\mathrm{d}\Phi}{\mathrm{d}x}\right) \tag{6.1}$$

6.3 Compact Ultrathin Linear Graded Index Metasurface Lens for Beam Steering and Gain Enhancement

In this section, a low-profile planar LGIMS lens is presented. A wide beam steerable high-gain low-profile antenna is designed by placing LGIMS over MPA radiator. Direction control of radiation pattern of the radiator using amplitude and phase-modulated MS is presented. The designed LGIMS lens causes beam steering when moved along negative parallel radiator axis and causes minor gain enhancement when moved along positive parallel radiator axis [14]. The prototype antenna is fabricated and characterized experimentally.

6.3.1 Design of Planar Single-Layer Linear Graded Index MS Lens

A planar modulated MS causes phase correction using localized phase shifters made up of UCs in the desired band. An MDCRR is used as an FUC to design linear GIMS, where FUC will act as a localized phase shifter. The FUC with dimensions and boundary conditions is shown in Figure 6.2(a) and (b), respectively.

The proposed MDCRR UC is designed by modifying a double CRR by adding capacitive slots to achieve a compactness of about 30 per cent. This MDCRR UC is complementary to DASR UC [15]. The simulated transmission and reflection characteristics with magnitude and phase plots are shown in Figure 6.3(a) and (b), respectively. Simulated 3-dB transmission band of 3.75 GHz from 8.75 GHz to 12.5 GHz with maximum transmission at 10.00 GHz is observed.

The group of modulated FUCs arranged in a sequence, as shown in Figure 6.4, is called super cell. The quantity "ΔD" ($\Delta D = 0.2$ mm) is simultaneous change in the dimension of outer mean ring radius "R" and inner mean ring radius "r" for next modulated FUC. The dimension change "ΔD" is constant for complete arrangement as shown in Figure 6.4. A data base of transmission magnitude and phase for several number of UCs is created for continuous positive and negative "ΔD" Several continuous dimensional changes carried out are simulated using CST microwave studio by defining boundary conditions as shown in Figure 6.2(b). It is observed that the change in transmission magnitude and phase is gradual and continuous.

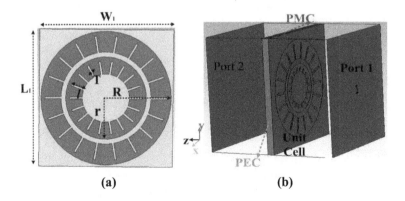

(a)   (b)

FIGURE 6.2

(a) FUC geometry and (b) simulation boundary condition. Dimensions are $L_1 = 14$ mm, $W_1 = 14$ mm, $R = 4$ mm, $r = 3$ mm, $l = 1$ mm and $T = 0.175$ mm.

Source: Copyright/used with permission of/courtesy of IEEE.

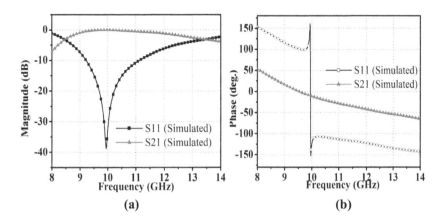

FIGURE 6.3
Simulated transmission and reflection coefficient. (a) Magnitude plot and (b) phase plot.

Source: Copyright/used with permission of/courtesy of IEEE.

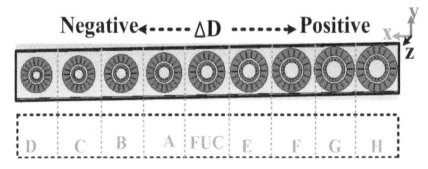

FIGURE 6.4
Modulated FUC arrangement, super cell.

Source: Copyright/used with permission of/courtesy of Wiley.

The transmission magnitude and phase variation of some of the optimum UC combinations for positive "ΔD" and negative "ΔD" are plotted in Figure 6.5 (a), Figure 6.5(b) and Figure 6.6(a) and Figure 6.6(b), respectively. Similar types of GIMS arrangement are also reported in [10].

The change in dimensions of modulated FUC changes surface current distribution, which in turn changes transmission magnitude and transmission phase, resulting in change in refractive index. It is observed that positive "ΔD" causes negative "ΔD" and negative "ΔD" causes positive "ΔD" where "ΔD" is the change intransmission phase.

The positive "ΔD" and negative "ΔD" cause a decrease in transmission magnitude, where rate of change in transmission magnitude is much higher for negative "ΔD". The change in transmission magnitude and transmission phase causes change in refractive index.

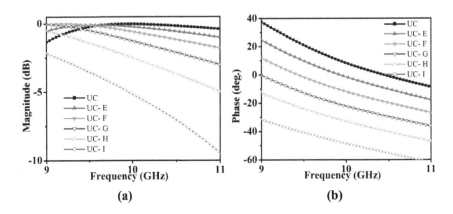

FIGURE 6.5
Simulated transmission coefficient of UC combinations for positive "ΔD". (a) Magnitude plot and (b) phase plot.

Source: Copyright/used with permission of/courtesy of Wiley.

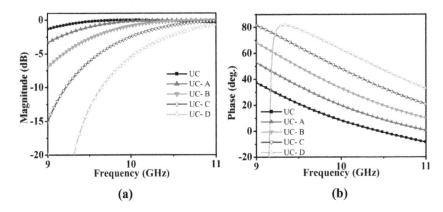

FIGURE 6.6
Simulated transmission coefficient of UC combinations for negative "ΔD". (a) Magnitude plot and (b) phase plot.

Source: Copyright/used with permission of/courtesy of Wiley.

The variation of refractive index for all optimum UCs having positive and negative "ΔD" is given in Table 6.1. It is observed that refractive index changes with respect to change in location of UC along parallel radiator axis (X-axis). The change in refractive index "$\Delta \eta$" is positive along positive X-axis and negative along negative X-axis. Changing the location along X-axis on the super cell in XY-plane changes transmission phase and magnitude gradually, resulting in gradual change in refractive index. Hence, the proposed super cell is a graded index super cell.

An MS lens is designed by placing six rows of super cell combination along Y-axis in XY-plane in linear orientation as shown in the fabricated prototype in

TABLE 6.1

Calculated Refractive Index of all Modulated FUCs

UCs	Refractive index	UCs	Refractive index
FUC	−0.96	FUC	−0.96
A	−1.98	E	−0.021
B	−3.37	F	0.124
C	−5.85	G	0.381
D	−8.45	H	2.85

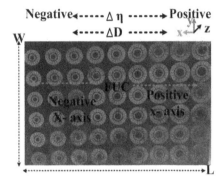

FIGURE 6.7
Fabricated prototype of LGIMS lens with refractive index profile distribution. Dimensions are $L = 126$ mm and = 84 mm.

Source: Copyright/used with permission of/courtesy of Wiley.

Figure 6.7. Hence, the designed MS lens is called as LGIMS lens. The LGIMS lens has two types of refractive index regions with respect to FUC, positive " $\Delta\eta$ " region along a positive X-axis and negative " $\Delta\eta$ " region along a negative X-axis, as shown in Figure 6.7. The LGIMS lens is fabricated on RT/duroid 5880 substrate having permittivity 2.2 and thickness $0.025\lambda_o$, making LGIMS lens ultrathin.

6.3.2 Design of a Beam Steerable High-Gain Antenna

A wide beam steerable high-gain antenna is designed by placing the ultrathin LGIMS lens over an MPA radiator fabricated on RT/duroid 5880 substrate operating at 10 GHz. The height of LGIMS lens over MPA is optimized to get maximum matching by varying "h". The designed prototype of the wide beam steerable high-gain LGIMS lens antenna is shown in Figure 6.8.

The variation in reflection coefficient of LGIMS lens antenna with lens height "h" variation is plotted in Figure 6.9. The optimum lens height "$h_{opt} = 12.30$ mm" is considered for achieving maximum matching. The simulated and measured reflection coefficients of MPA and LGIMS lens antenna are plotted in Figure 6.10. Both are found to be in good agreement with small variation due to loading effect and alignment error.

FIGURE 6.8
Fabricated prototype of the compact LGIMS lens antenna.

Source: Copyright/used with permission of/courtesy of Wiley.

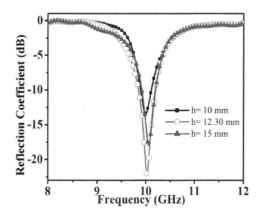

FIGURE 6.9
Reflection coefficient with LGIMS lens height variation over MPA.

Source: Copyright/used with permission of/courtesy of Wiley.

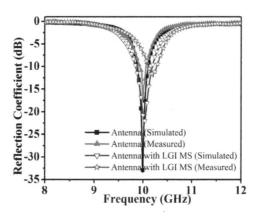

FIGURE 6.10
Measured and simulated reflection coefficients of MPA and LGIMS lens antenna.

Source: Copyright/used with permission of/courtesy of Wiley.

6.3.3 Measured Results

The normalized radiation characteristics of both MPA and LGIMS lens antennas are measured in an anechoic chamber and H-plane (ZX-plane) simulated and measured radiation characteristics are plotted in Figure 6.11. A broadside gain enhancement of 6.32 dB is observed. Similar simulated and measured radiation characteristics are observed for E-plane (YZ-plane).

To steer the beam of the LGIMS lens antenna, the lens is moved along the negative X-axis in steps of 15 mm. The measured H-plane radiation characteristics of the LGIMS lens antenna for parameter "P" variation, where "P" is the movement of the LGIMS lens along the X-axis (see Figure 6.7), is plotted in Figure 6.12. It is observed that continuous variation of "P" in steps of 15 mm along the negative X-axis causes steering of the main beam of MPA by 9°, 20° and 32°, corresponding to "P" variation of –15 mm, –30 mm and –45 mm, respectively, forming a conical plane with a vertex angle of "64°" having peak gain variation from 11.07 dB to 13.50 dB, indicating gain enhancement of 6.32 dB to 8.75 dB, respectively, as shown in Figure 6.12. The measured E-plane radiation pattern also has similar characteristics. The change in refractive index "$\Delta\eta$" is negative along the negative X-movement on LGIMS lens, due to which the radiated beam of MPA moves away from the broadside direction, resulting in beam steering.

The simulated H-plane radiation characteristics due to positive X-movement of "P" on the LGIMS lens is plotted in Figure 6.13, indicating progressive broadside gain enhancement. The positive X-movement of the LGIMS lens causes positive "$\Delta\eta$", due to which the radiated beam moves towards broadside direction, resulting in significant increase in broadside gain without steering of the main radiated beam.

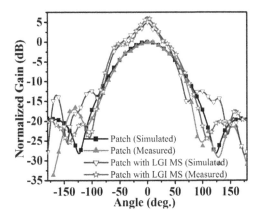

FIGURE 6.11
Measured and simulated H-plane (ZX) radiation characteristics of MPA and LGIMS lens antenna.

Source: Copyright/used with permission of/courtesy of Wiley.

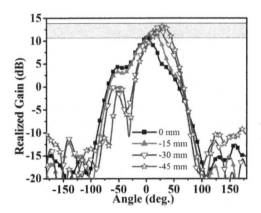

FIGURE 6.12
Measured H-plane (ZX-plane) radiation characteristics of MPA and LGIMS lens antenna with "P" variations (negative).

Source: Copyright/used with permission of/courtesy of Wiley.

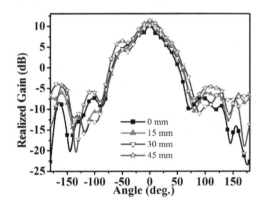

FIGURE 6.13
Simulated H-plane (ZX-plane) radiation characteristic of MPA and LGIMS lens antenna with "P" variations (positive).

Source: Copyright/used with permission of/courtesy of Wiley.

The characteristics of the proposed antenna are compared with several similar reported antennas in Table 6.2 and are found to be more compact, offering high-gain with wide beam steering conical plane apex angle.

A compact ultrathin linear GIMS lens is successfully designed. A compact wide angle beam steering high-gain antenna with a conical plane vertex angle of 64° having peak gain variation from 11.07 dB to 13.50 dB, indicating gain enhancement of 6.32 dB to 8.75 dB with significant reduction in side lobe level, is achieved by placing LGIMS lens over an MPA radiator.

TABLE 6.2

Comparison of the Proposed Antenna with Other Reported Antennas

Ref.	Operating Frequency (GHz)	Lens Dimension ("λ" is wavelength at centre operating frequency)	Beam Steering (degree)	Gain Enhancement (dB)
[16]	9.9 to 10.1	4λ × 4λ × 0.034λ	——	——
[17]	7.0 to 10.0	5.10λ × 1.75λ × 0.14λ	−10 to 18.9	——
[18]	8.8 to 10.7	9.7λ × 0.65λ × 0.077λ	4.8 to 37.2	——
This work	**9.68 to 10.30**	**4.2λ × 2.82λ × 0.025λ**	**0 to 32** (apex angle 64°)	**6.32 to 8.75**

6.4 Radial/Angular Graded Index Metasurface Lens for Beam Steering and Gain Enhancement

In this section, an ultrathin highly transmissive RGIMS lens is proposed and a high-gain wide beam steerable low-profile lens antenna is designed by placing RGIMS lens on the top of MPA radiator operating at 10.10 GHz. The direction and gain control of radiated beam of microwave radiator using modulated MS having MDCRR as FUC is presented. The RGIMS lens can continuously steer the radiated beam of MPA radiator within a 2-dB gain variation band by radial movement of RGIMS [19].

6.4.1 Design of a Radial Graded Index Metasurface (RGIMS)

The proposed MDCRR UC (FUC) and the three negatively modulated UCs—A, B and C—with negative "ΔD", discussed in Section 6.3.1, are considered for the design of a radial GIMS. The simulated transmission magnitude and transmission phase of these UCs are plotted in Figure 6.6(a) and Figure 6.6 (b), respectively. It is observed that negative "ΔD" causes negatively graded transmission magnitude change and positively graded transmission phase change. The refractive index of these modulated MDCRR UCs is shown in Table 6.1. It is observed that negative "ΔD" causes negative graded change in the refractive index.

These modulated UCs are arranged in a nearly constant circular phase distribution such that all the cell elements in the same circular region will have same transmission magnitude and transmission phase as shown in Figure 6.14. Similar type of GIMS design is also proposed in [20–21]. Angular or radial traversal in the region will cause graded change in transmission magnitude and phase, resulting in graded refractive index profile. The nearly uniform circular phase profile arrangement is shown in Figure 6.14. Here 1 = FUC, 2 = A, 3 = B and 4 = C (FUC with negatively modulated UCs).

			6	6	6	6				
	6	6	5	5	5	5	6	6		
6	5	5	4	4	4	4	5	5	6	
6	5	4	3	3	3	3	4	5	6	
5	4	3	3	2	2	3	3	4	5	
5	4	3	2	1	1	2	3	4	5	
5	4	3	2	1	1	2	3	4	5	
5	4	3	3	2	2	3	3	4	5	
6	5	4	3	3	3	3	4	5	6	
6	5	5	4	4	4	4	5	5	6	
	6	6	5	5	5	5	6	6		
			6	6	6	6				

FIGURE 6.14
The nearly uniform circular phase distribution profile arrangement.

Source: Copyright/used with permission of/courtesy of IEEE.

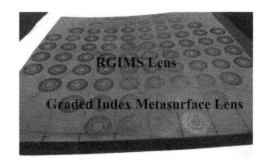

FIGURE 6.15
The fabricated RGIMS lens.

Source: Copyright/used with permission of/courtesy of IEEE.

This arrangement of UCs results in graded change in transmission magnitude and transmission phase along the radius of the circular plane ($\frac{d\Phi}{dr}$), where "Φ" is phase profile of the UC. This arrangement also results in graded change in refractive index along the radius of circular plane ($\frac{d\eta}{dr}$). Due to this arrangement, the designed MS is called RGIMS. The RGIMS is fabricated on top of RT/duroid 5880 substrate having permittivity 2.2 and thickness 0.762 mm (0.025λ) using photolithography. The fabricated RGIMS is shown in Figure 6.15.

6.4.2 Design of an RGIMS Lens Antenna and Measured Results

The fabricated RGIMS is placed on top of the MPA radiator operating at 10.10 GHz at an optimum height to provide maximum matching. The height of RGIMS over MPA is optimized using CST microwave studio and found to be 12.50 mm. A high-gain wide beam steerable low-profile lens antenna is designed by placing the fabricated RGIMS on top of the MPA radiator operating at 10.10 GHz at an optimum height "h = 12.50 mm". The fabricated prototype of the RGIMS lens antenna is shown in Figure 6.16. The reflection coefficient of RGIMS lens antenna is measured using VNA and plotted in Figure 6.17. The reflection coefficient of MPA is disturbed by loading of the RGIMS. This error may be due to alignment problem also. The H-plane radiation characteristics of RGIMS are measured in an anechoic chamber and plotted in Figure 6.18. A broadside gain enhancement of 6.5 dB, as compared to MPA broadside gain, is observed. The designed MPA operating at 10.15 GHz shows measured broadside gain of 4.5 dB.

To steer the radiated beam of MPA, the RGIMS is moved along the X-axis in steps of 14 mm with respect to point $P(X = 0$ mm, $Y = 0$ mm), as shown in Figure 6.16. The measured reflection coefficient due to movement of RGIMS along the positive X-axis in steps of 14 mm is plotted in Figure 6.17. A good agreement between measured results is obtained. Similar reflection coefficient characteristics are observed due to the negative X-axis movement of RGIMS in steps of 14 mm.

The measured H-plane radiation characteristics due to movement of RGIMS along the positive X-axis in steps of 14 mm are plotted in Figure 6.18. It is observed that the positive X-axis movement of RGIMS in steps of 14 mm steers the main radiated beam of MPA by 10°, 20° and 32° with a measured peak gain of 12.48 dB, 13.82 dB and 14.71 dB, indicating gain enhancement of 7.98 dB, 9.32 dB and 10.21 dB, corresponding to movement of 14 mm, 28 mm and 42 mm, respectively.

The H-plane radiation characteristics of RGIMS lens antenna due to the negative X-axis movement of the RGIMS in steps of 14 mm with respect to point "$P (X = 0$ mm, $Y = 0$ mm)", as shown in Figure 6.16, are measured. It is observed that negative X-axis movement of RGIMS in steps of 14 mm steers the main re-radiated beam of MPA by −10°, −20° and −32° with measured peak gain of 12.45 dB, 13.79 dB and 14.41 dB, indicating gain enhancement of 7.95 dB,

RGIMS Lens Antenna

FIGURE 6.16
The fabricated prototype of RGIMS lens antenna.

Source: Copyright/used with permission of/courtesy of IEEE.

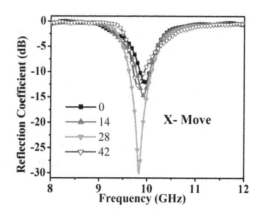

FIGURE 6.17
Measured reflection coefficient of RGIMS lens antenna and movement of RGIMS along positive X-axis in steps of 14 mm.

Source: Copyright/used with permission of/courtesy of IEEE.

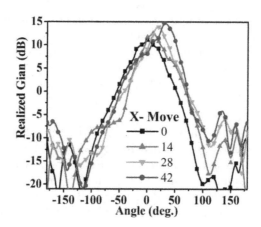

FIGURE 6.18
Measured radiation pattern due to positive X-axis movement of RGIMS in steps of 14 mm.

Source: Copyright/used with permission of/courtesy of IEEE.

9.29 dB and 9.91 dB corresponding to movement of −14 mm, −28 mm and −42 mm, respectively, as shown in Figure 6.19. Similar radiation characteristics are observed for positive and negative Y-axis movement of RGIMS in steps of 14 mm.

It can also be observed from Figures 6.18 and 6.19 that RGIMS lens can steer the main radiated beam of MPA in a conical plane of vertex angle 64° with maximum broadside gain of 14.71 dB and minimum broadside gain of 11.07 dB in 32° and 0° directions, respectively. RGIMS lens can continuously steer the radiated beam of MPA with significant high gain.

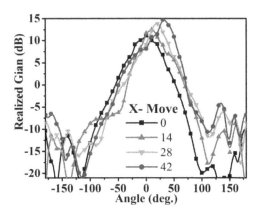

FIGURE 6.19
Measured radiation pattern due to negative *X*-axis movement of RGIMS in steps of 14 mm.

Source: Copyright/used with permission of/courtesy of IEEE.

A low-profile radial GIMS with modulated transmission magnitude and phase characteristics is designed first. A high-gain low-profile continuous wide beam steerable RGIMS lens antenna with beam steering in a conical plane of vertex angle 64° by radial movement in steps of 14 mm is successfully presented.

6.5 Wide Angle Beam Steerable High-Gain Flat Top Beam Antenna Using a Graded Index Metasurface

In this section, development of a flat top beam (FTB) high-gain wide beam steerable low-profile MS lens antenna using GIMS lens on the top of an MPA radiator operating at 10.10 GHz is presented. An ultrathin LGIMS and angular graded index metasurface (AGIMS) lens with wide angular stability is designed for near-field phase transformation of radiated beam by microwave source radiator [22]. The simulated and measured results are found to be in good agreement.

6.5.1 Design of the Transparent Unit Cell

The MS is designed using the sub-wavelength double-sided double circular ring resonator (DSDCRR) as the FUC. DSDCRR consists of two CRRs) printed on two sides of RT/duroid 5880 substrate having permittivity 2.2 and thickness 0.762 mm (0.025λ). The outer and inner CRRs on front and backside of dielectric substrate differ in radial dimension of outer and inner CRRs by 0.50 mm and 0.25 mm, respectively. The detailed geometry of the DSDCRR

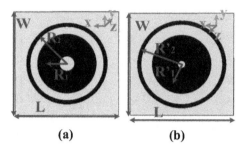

(a)   **(b)**

FIGURE 6.20

FUC geometry. (a) Top view and (b) back view. Dimensions are $R_1 = 2.25$ mm, $R_1' = 2.00$ mm, $R_2 = 4.75$ mm, $R_2' = 5.00$ mm and $L = 12$ mm and $W = 12$ mm.

Source: Copyright/used with permission of/courtesy of IEEE.

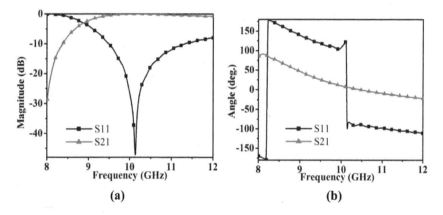

(a)   **(b)**

FIGURE 6.21

Simulated transmission and reflection coefficient of DSDCRR FUC. (a) Magnitude plot and (b) phase plot.

Source: Copyright/used with permission of/courtesy of IEEE.

UC with front and back views with dimensions is shown in Figure 6.20(a) and (b), respectively. This UC is considered as the UC. The DSDCRR UC is simulated using CST microwave studio, having PEC boundary condition along the X-axis and PMC boundary condition along the Y-axis. The UC is excited by the normal incidence of EM waves with two WG ports along the Z-axis.

The simulated transmission and reflection characteristics of DSDCRR FUC with magnitude and phase characteristics are plotted in Figure 6.21(a) and (b), respectively. A simulated wide 3-dB transmission band of more than 3.15 GHz with simulated transmission delay time of 60 ps at 10.14 GHz is observed.

The DSDCRR FUC is modulated by a dimension change of "ΔR", where "ΔR" is change in radial dimension of inner and outer CRRs on top and bottom surfaces. Several UCs with corresponding positive and negative "ΔR" values are simulated using CST microwave studio using PEC and PMC boundary conditions, and a data base of simulated transmission magnitude and phase with transmission delay time is created.

Some of the simulated UCs having high transmission magnitude, uniform transmission phase change and close transmission delay time are selected. The transmission magnitude and transmission phase plot of selected UCs with positive and negative radial dimension change "ΔR" are plotted in Figure 6.22 and Figure 6.23, respectively. It is observed that negative "ΔR" causes very small decrease in transmission magnitude and positive

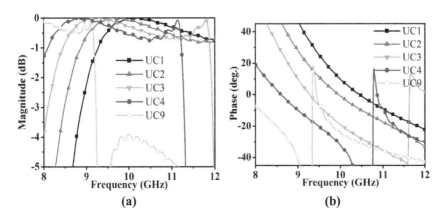

FIGURE 6.22
Simulated transmission coefficient of UC combinations for negative "ΔR". (a) Magnitude plot and (b) phase plot.

Source: Copyright/used with permission of/courtesy of IEEE.

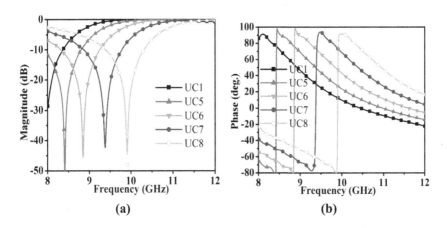

FIGURE 6.23
Simulated transmission coefficient of UC combinations for positive "ΔR". (a) Magnitude plot and (b) phase plot.

Source: Copyright/used with permission of/courtesy of IEEE.

TABLE 6.3

Calculated Refractive Index and Transmission Delay Time of All Modulated FUCs

UCs	Refractive Index	Delay Time (ps)	UCs	Refractive Index	Delay Time (ps)
FUC	−0.842	59.12	FUC	−0.842	59.12
UC1	−1.799	71.38	UC5	0.297	51.10
UC2	−2.766	83.06	UC6	0.561	45.02
UC3	−4.067	89.67	UC7	0.953	43.39
UC4	−5.863	92.45	UC8	1.032	39.57

"ΔR" causes very fast decrease in transmission magnitude as shown in Figure 6.22(a) and (b), respectively.

Negative "ΔR" causes progressive decrease in transmission phase and positive "ΔR" causes a progressive increase in transmission phase as shown in Figure 6.22(b) and Figure 6.23(b), respectively. The corresponding refractive index and transmission delay time of modulated FUCs are given in Table 6.3. It is observed that positive "ΔR" causes positive change in refractive index and negative "ΔR" causes negative change in refractive index. Successive increase in transmission delay time is observed due to negative "ΔR" and successive decrease in delay time is observed due to positive "ΔR".

6.5.2 Design of the Linear Graded Index (LGIMS) Metasurface

The modulated FUC combinations, UC1, UC2, . . ., UC7 and UC8, are arranged along the X-axis in the XY-plane as shown in Figure 6.24. This arrangement of FUCs is called a super cell. Considering FUC as the centre, it is observed that dimension change "ΔR" is negative along negative X-axis and "ΔR" is positive along positive X-axis, as shown in Figure 6.24.

Due to the earlier arrangement shown in Figure 6.24, the negative refractive index region occurs along the negative X-axis and positive refractive index region occurs along the positive X-axis. A positive increased transmission delay time region is observed along the negative X-axis and negative increased transmission delay time region along the positive X-axis with respect to FUC as shown in Figure 6.24. The change in refractive index region and transmission delay time region is due to modulation of the dimensions of FUC.

An MS is designed by placing six rows of super cell along the Y-axis. Due to this, the designed surface has modulated FUCs placed along the X-axis. As modulated FUC arrangement is placed on a plane along a line where the change in transmission magnitude, transmission phase and refractive index is graded along the X-axis, the designed MS is called LGIMS. The LGIMS is fabricated using photolithography on RT/duroid 5880 substrate having permittivity 2.2 and thickness 0.762 mm (0.025λ). The fabricated prototype of LGIMS is shown in Figure 6.25.

The FUC is indicated by the black dotted line. The point "P" on the LGIMS is a reference point and considered as origin $P(X = 0$ mm, $Y = 0$ mm) as shown in Figure 6.25.

FIGURE. 6.24
The modulated DSDCRR FUC arrangement and super cell.

Source: Copyright/used with permission of/courtesy of IEEE.

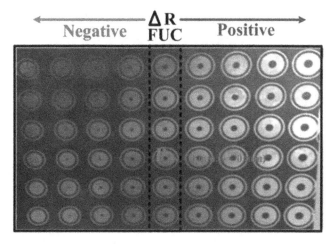

FIGURE 6.25
Fabricated DSDCRR LGIMS prototype.

Source: Copyright/used with permission of/courtesy of IEEE.

6.5.3 Design of the Angular Graded Index Metasurface

A nearly uniform circular graded index region is created by considering the proposed FUC as shown in Figure 6.20(a) and (b), hereafter denoted as "1". Three modulated forms of 1 (FUC) are created having constant negative corresponding radial dimension change "ΔR" named as 2, 3 and 4. The geometrical dimension of 1 is same as FUC, 2 is same as UC1, 3 is same as UC2 and 4 is same as UC3 proposed in Section 6.5.2. The UCs 1, 2, 3 and 4 are arranged in a way so as to design nearly uniform circular phase distribution as shown in Figure 6.26. It is observed that the dimensional change "ΔR" is negative along any radial direction with respect to 1 (FUC). Due to this arrangement, negative refractive index region occurs along positive or negative radial direction with respect to 1 (FUC). This arrangement generates negative graded refractive index region and positive increased transmission delay time region along radial directions. Due to this, the designed MS is called RGIMS/AGIMS. The AGIMS surface is fabricated on RT/duroid 5880 substrate having permittivity 2.2 and thickness 0.762 mm (0.025λ). The

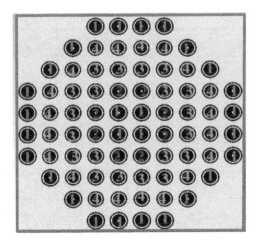

FIGURE 6.26
The UC arrangement for AGIMS.

Source: Copyright/used with permission of/courtesy of IEEE.

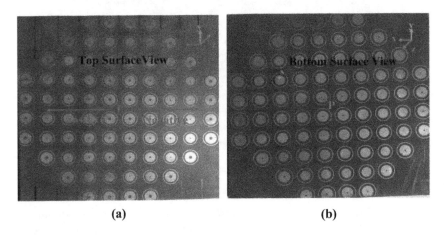

(a)   **(b)**

FIGURE 6.27
Fabricated AGIMS. (a) Top view and (b) bottom view.

Source: Copyright/used with permission of/courtesy of IEEE.

fabricated AGIMS prototype is shown in Figure 6.27(a) and Figure 6.27(b) as top and bottom surface views, respectively. The point "P" is the reference point and is considered as origin P ($X = 0$ mm, $Y = 0$ mm).

6.5.4 Design of the LGIMS Lens Antenna and Measurement Results

The designed low-profile ultrathin LGIMS is placed at an optimum height on the top of the MPA radiator. The optimum height is obtained by analysing reflection coefficient variation of MPA due to LGIMS height variations as shown in Figure 6.28.

It is observed that maximum matching is achieved at 10 mm. The LGIMS and MPA operating at 10 GHz are fabricated on RT/duroid 5880 substrate having permittivity 2.2 and thickness 0.762 mm. The fabricated LGIMS is placed over MPA at an optimum height "h = 10 mm". The prototype LGIMS lens antenna is shown in Figure 6.29. The reflection coefficient of MPA and LGIMS lens antenna is measured and compared in Figure 6.30. A good matching is observed. The radiation characteristics of MPA and LGIMS lens antenna is measured in an anechoic chamber. The measured H-plane radiation characteristics of LGIMS lens antenna and movement of LGIMS lens along negative X-axis is plotted in Figure 6.31. A broadside gain enhancement of 8.39 dB as compared to MPA is observed with maximum measured gain of 13.24 dB as shown in Figure 6.31. The MPA radiator has measured gain of 4.85 dB.

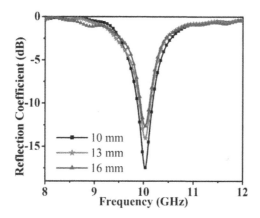

FIGURE 6.28
Simulated reflection coefficient due to height variation of LGIMS.

Source: Copyright/used with permission of/courtesy of IEEE.

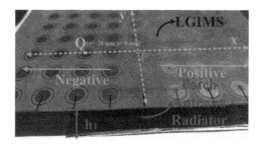

FIGURE 6.29
Fabricated LGIMS lens antenna prototype.

Source: Copyright/used with permission of/courtesy of the IEEE.

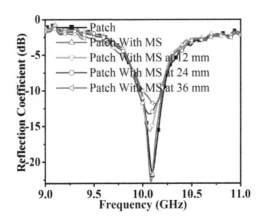

FIGURE 6.30
Measured reflection coefficient of MPA and LGIMS at various *X*-axis movements of LGIMS.

Source: Copyright/used with permission of/courtesy of IEEE.

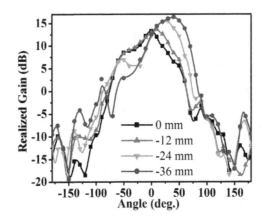

FIGURE 6.31
Measured H-plane radiation characteristic of LGIMS lens antenna due to various moves along negative *X*-axis.

Source: Copyright/used with permission of/courtesy of IEEE.

To steer the beam of MPA radiator with gain enhancement, LGIMS lens is moved along the *X*-axis in steps of 12 mm. The measured reflection coefficient of LGIMS lens antenna due to movement of LGIMS along positive *X*-axis in steps of 12 mm is shown in Figure 6.30. A good matching is observed. Similar type of matching is observed for the negative *X*-axis movement of LGIMS lens.

The measured H-plane (XZ-plane) radiation characteristic of LGIMS lens antenna due to movement of LGIMS along the negative *X*-axis with respect to point "*P*(*X* = 0 mm, *Y* = 0 mm)", as shown in Figure 6.29 in steps of 12 mm, is plotted in Figure 6.31. It is observed that movement of LGIMS causes steering of the main beam of MPA radiator by 10°, 30° and 42°, corresponding to LGIMS movement of −12 mm, −24 mm and −36 mm, respectively, forming a conical plane with a vertex angle of "84°" having peak gain variation from 13.25 dB to 16.37 dB, indicating gain enhancement of 7.35 dB to 10.70 dB, respectively, as shown in Figure 6.31. The radiation characteristics due to movement of LGIMS along the positive X-axis with respect to point "P(X = 0 mm, Y = 0 mm)" in steps of 12 mm are plotted in Figure 6.32. It is observed that the positive X-axis movement of LGIMS causes gain enhancement by 1.5 dB as shown in Figure 6.32.

The effect of angular movement of LGIMS about any arbitrary point is studied next. The radiation pattern of MPA due to angular movement of LGIMS is analysed at point "Q(X = −24 mm, Y = 0 mm), as shown in Figure 6.29. The LGIMS is moved by −24 mm along the negative X-axis such that point "Q" is aligned with the MPA radiator's centre, so that the radiated beam of the LGIMS lens antenna should be steered in the direction of 20°. Next, the LGIMS is rotated in clockwise direction in steps of 30°. The measured H-plane (XZ-plane) radiation characteristic due to this angular movement is plotted in Figure 6.33. It is observed that angular clockwise movement of LGIMS by 0°, 30° and 60° causes beam steering in the direction of 20°, 10° and 0° with measured peak gain of 15.56 dB, 14.68 dB and 13.48 dB, respectively, as shown in Figure 6.33(a). Further movement of LGIMS in clockwise direction by 90°, 120° and 150° causes beam steering of MPA radiator in the direction of −10°, −15° and −25° with measured peak gain of 12.50 dB, 13.43 dB and 14.40 dB, respectively, as shown in Figure 6.33(b). Further movement of LGIMS in clockwise direction by 180°, 210° and 240° causes beam steering in the direction of −35°, −25° and −10° with measured peak gain of 15.15 dB, 14.35 dB and 13.24 dB, respectively, as shown in Figure 6.33(c). Further movement of LGIMS by 270°, 300° and 330° in clockwise direction causes beam steering in the direction of 0°, 10°

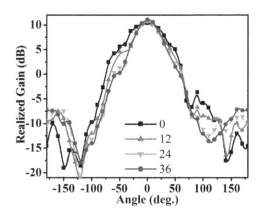

FIGURE 6.32
Measured H-plane radiation characteristic due to positive X-axis movement of LGIMS.

Source: Copyright/used with permission of/courtesy of IEEE.

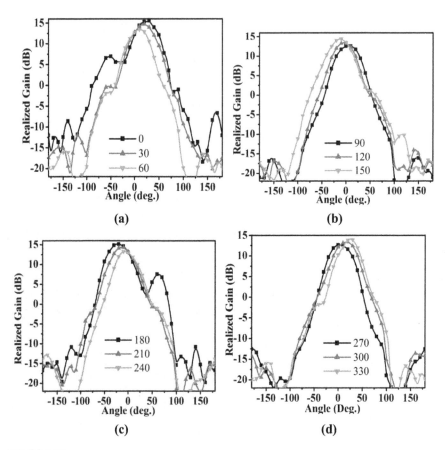

FIGURE 6.33
Measured H-plane radiation characteristic of LGIMS lens antenna due to angular movement of LGIMS.

Source: Copyright/used with permission of/courtesy of IEEE.

and 15° with measured peak gain of 12.66 dB, 13.40 dB and 14.24 dB, respectively as shown in Figure 6.33(d).

It can be concluded from the measured results in Figure 6.33 that the source broadside beam generated due to linear movement of LGIMS along the negative X-axis can be steered in any direction within a conical plane of vertex angle about $2\theta_L$ with maximum measured gain variation of 3 dB (for $X = -24$ mm and $Y = 0$ mm it is 12.50 dB to 15.56 dB), where θ_L is beam steering angle due to linear movement of the LGIMS along the X-axis. The linear movement of LGIMS over MPA radiator changes radiated phase profile of MPA to nearly equal planar phase distribution. The angular movement of LGIMS causes nearly equal circular phase distribution profile, resulting in beam steering within a conical plane with vertex angle of about two times the beam steering angle due to linear movement of LGIMS.

It is also observed from Figure 6.31 and Figure 6.33 that the HPBW of radiated beam by LGIMS lens antenna is very small. This small HPBW can further be improved by changing the planar phase distribution profile radiated by LGIMS lens antenna to nearly circular equal phase distribution profile by placing a phase-correcting surface in the near-field region of LGIMS lens antenna.

6.5.5 Design of the Flat Top Beam Antenna

A flat top radiated beam GIMS lens antenna is designed using LGIMS and AGIMS. The HPBW of radiated beam of LGIMS lens antenna is improved by near-field phase transformation using AGIMS. The designed AGIMS/RGIMS in Figure 6.27 is nearly equal circular phase distribution profile MS. The AGIMS is placed on top of the LGIMS lens antenna at an optimum height of "h_2". The height of AGIMS over the LGIMS lens antenna is optimized using CST microwave studio software and maximum matching is observed at "$h_2 = 6$ mm". A wide beam steerable high-gain FTB antenna with improved HPBW is designed by placing fabricated AGIMS at an optimum height of "$h_2 = 6$ mm" on top of the LGIMS lens antenna. The prototype antenna is shown in Figure 6.34. The optimized height of LGIMS "h_1" and AGIMS "h_2" causes maximum matching.

The performance of an FTB antenna is characterized by measuring radiation pattern in an anechoic chamber. The radiation pattern of FTB antenna due to linear and angular movement of LGIMS and AGIMS is measured. The H-plane radiation pattern of FTB antenna due to linear movement of LGIMS in steps of 12 mm along negative X-axis keeping AGIMS fixed is plotted in Figure 6.35. It is observed that the radiation pattern is flat in nature having 1-dB maximum gain variation band of 30° from −15° to 15° with maximum measured gain of 13.36 dB in 0° directions. It is also observed that the radiation pattern is maximum flat in nature, having 1-dB maximum gain variation band of 55° from −5° to 50° with maximum measured gain of 16.03 dB in 15° directions due to LGIMS movement by −36 mm [22]. Linear movement of LGIMS keeping RGIMS fixed causes beam steering of MPA with improved radiated gain and beam broadening band with 1-dB maximum gain variation. The AGIMS on the top of LGIMS lens antenna causes local amplitude and phase modification due to near-field effect of AGIMS, resulting into beam broadening. The small ripple in the measured H-plane radiation pattern of FTB antenna is due to alignment problem.

FIGURE 6.34
The prototype FTB GIMS lens antenna.

Source: Copyright/used with permission of/courtesy of IEEE.

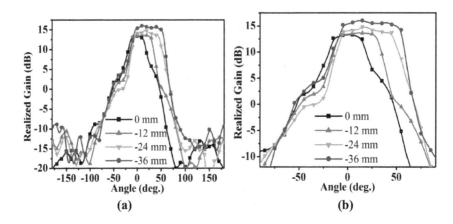

FIGURE 6.35
(a) Measured H-plane radiation characteristics of FTB antenna due to linear movement of LGIMS and (b) zoomed view of the same pattern.

Source: Copyright/used with permission of/courtesy of IEEE.

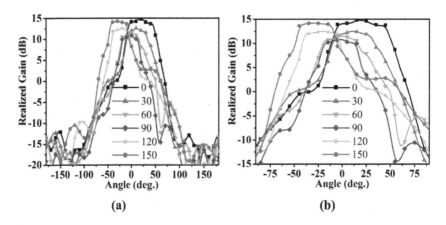

FIGURE 6.36
(a) Measured H-plane radiation characteristic of FTB antenna due to angular movement of LGIMS and (b) zoomed view of the pattern.

Source: Copyright/used with permission of/courtesy of IEEE.

To achieve wider gain performance, the LGIMS is moved to point "$Q(X = -24$ mm, $Y = 0$ mm)" and rotated in clockwise direction in steps of 30° keeping AGIMS fixed. The H-plane radiation characteristics due to this clockwise rotation are plotted in Figure 6.36. The angular movement of LGIMS of FTB antenna in clockwise direction in this condition results in FTB having 1-dB gain variation band of maximum 53° from about −5° to 48° with maximum measured gain of 14.82 dB in 15° directions due to 0° move of LGIMS. A 150° clockwise move of LGIMS results in FTB having 1-dB gain variation band of 53° from −48° to −8° with maximum gain of 14.25 dB in −30° direction. This results in continuous beam scanning from −48° to 48° with average measured gain of about 14 dB as shown in Figure 6.36. It is also observed that when LGIMS is moved to point $R(X = -36$ mm, $Y = 0$ mm) and rotated in clockwise direction in steps of 30°, a wide beam scanning from −53° to +53° with maximum gain of 15.56 dB in 42° direction is observed. This results in continuous beam scanning from −53° to +53°, with average measured gain of about 15 dB [22].

A compact high-gain wide beam steerable FTB antenna using GIMS lens on the top of MPA radiator as microwave source is successfully demonstrated. The designed FTB antenna has 1-dB gain variation band of 96° from −48° to −5° and −8° to −48° with maximum gain of 14.25 dB, resulting in continuous beam scanning from about−50° to 50° with maximum gain enhancement of about 10 dB [22].

6.6 Conclusion

In this chapter, GIMS lens is designed for X-band applications. Two types of GIMS are designed, LGIMS and AGIMS/RGIMS. The designed GIMS are used to steer the main radiated beam of MPA with gain enhancement. An FTB antenna having beam scanning ability from about −50° to +50° with nearly 1-dB gain variation band of about 96° from −48° to −5° and −8° to−48° with maximum gain of 14.25 dB is successfully designed.

In Chapter 7, the design and development of ultrathin compact microwave metamaterial absorber for S-, C-, X- and Ka-band applications is presented along with measured results on the fabricated prototypes.

References

1. Li, H., G. Wang, H. X. Xu, T. Cai, and J. Liang, "X-band phase-gradient metasurface for high-gain lens antenna application," *IEEE Transactions on Antennas and Propagation*, vol. 63, no. 11, pp. 5144–5149, Nov. 2015.
2. Ratni, B., W. A. Merzouk, A. de Lustrac, S. Villers, G. P. Piau, and S. N. Burokur, "Design of phase-modulated metasurfaces for beam steering in Fabry–Perot cavity antennas," *IEEE Antennas and Wireless Propagation Letters*, vol. 16, pp. 1401–1404, 2017.
3. Li, J., Q. Zeng, R. Liu, and T. A. Denidni, "Beam-tilting antenna with negative refractive index metamaterial loading," *IEEE Antennas and Wireless Propagation Letters*, vol. 16, pp. 2030–2033, 2017.

4. Ji, L. Y., Y. J. Guo, P. Y. Qin, S. X. Gong, and R. Mittra, "A reconfigurable partially reflective surface (PRS) antenna for beam steering," *IEEE Transactions on
   Antennas and Propagation*, vol. 63, no. 6, pp. 2387–2395, June 2015.
5. Li, T. and Z. N. Chen, "Control of beam direction for substrate-integrated waveguide slot array antenna using metasurface," *IEEE Transactions on Antennas and
   Propagation*, vol. 66, no. 6, pp. 2862–2869, June 2018.
6. Cheng, G., Y. M. Wu, J. X. Yin, N. Zhao, T. Qiang, and X. Lv, "Planar Luneburg
   lens based on the high impedance surface for effective Ku-band wave focusing," *IEEE Access*, vol. 6, pp. 16942–16947, 2018.
7. Larimore, Z., S. Jensen, A. Good, A. Lu, J. Suarez, and M. Mirotznik, "Additive
   manufacturing of Luneburg lens antennas using space-filling curves and
   fused filament fabrication," *IEEE Transactions on Antennas and Propagation*, vol.
   66, no. 6, pp. 2818–2827, June 2018.
8. Markovich, H., D. Filonov, I. Shishkin, and P. Ginzburg, "Bifocal Fresnel
   lens based on the polarization-sensitive metasurface," *IEEE Transactions on
   Antennas and Propagation*, vol. 66, no. 5, pp. 2650–2654, May 2018.
9. Katare, K., A. Biswas, and M. J. Akhtar, "Microwave beam steering of planar
   antennas by hybrid phase gradient metasurface structure under spherical
   wave illumination", *Journal of Applied Physics*, vol. 122, no. 23, p. 234901, 2017.
10. Afzal, M. U. and K. P. Esselle, "Steering the beam of medium-to-high gain
    antennas using near-field phase transformation," *IEEE Transactions on Antennas
    and Propagation*, vol. 65, no. 4, pp. 1680–1690, April 2017.
11. Zhou, Hai-Jin, You-Huo Huang, Bao-Hua Sun, and Qi-Zhong Liu, "Design and
    realization of a flat-top shaped-beam antenna array," *Progress in Electromagnetics
    Research*, vol. 5, pp. 159–166, 2008.
12. Zhang, Z. Y., N. W. Liu, S. Zuo, Y. Li, and G. Fu, "Wideband circularly polar-
    ized array antenna with flat-top beam pattern," *IET Microwaves, Antennas &
    Propagation*, vol. 9, no. 8, pp. 755–761, June 5, 2015.
13. Monavar, F. M., S. Shamsinejad, R. Mirzavand, J. Melzer, and P. Mousavi,
    "Beam-steering SIW leaky-wave subarray with flat-topped footprint for 5G
    applications," *IEEE Transactions on Antennas and Propagation*, vol. 65, no. 3,
    pp. 1108–1120, March 2017.
14. Singh, Amit K., Mahesh P. Abegaonkar, and Shiban K. Koul, "Compact ultra-
    thin linear graded index metasurface lens for beam steering and gain enhance-
    ment," *International Journal of RF and Microwave Computer-Aided Engineering*,
    pp. 202–210, March 2020.
15. Singh, Amit K., Mahesh P. Abegaonkar, and Shiban K. Koul, "High-gain and
    high-aperture-efficiency cavity resonator antenna using metamaterial super-
    strate," *IEEE Antennas and Wireless Propagation Letters*, vol. 16, pp. 2388–2391,
    June 2017.
16. Feng, Pan, Xing Chen, and Kama Huang, "High performance resonant cavity
    antenna with non-uniform metamaterial inspired superstrate," *International
    Journal of RF and Microwave Computer-Aided Engineering*, vol. 27, no. 7, Sep. 2017.
17. Chen, H. et al., "Wideband frequency scanning SSPP planar antenna based
    on transmissive phase gradient metasurface," *IEEE Antennas and Wireless
    Propagation Letters*, no. 99, p. 1, Feb. 2018.
18. Fan, Y. et al., "Frequency scanning radiation by decoupling spoof surface plas-
    mon polaritons via phase gradient metasurface," *IEEE Transactions on Antennas
    and Propagation*, vol. 66, no. 1, pp. 203–208, Jan. 2018.

19. Singh, A. K., Mahesh P. Abegaonkar, and Shiban K. Koul, "Radial graded index metasurface lens for beam steering and gain enhancement," *2018 International Symposium on Antennas and Propagation (ISAP)*, Busan, pp. 1–2, 2018.

20. Afzal, M. U. and K. P. Esselle, "Steering the beam of medium-to-high gain antennas using near-field phase transformation," *IEEE Transactions on Antennas and Propagation*, vol. 65, no. 4, pp. 1680–1690, April 2017.

21. Katare, Kranti Kumar, Animesh Biswas, and M. Jaleel Akhtar, "Microwave beam steering of planar antennas by hybrid phase gradient metasurface structure under spherical wave illumination," *Journal of Applied Physics*, vol. 122, no. 23, p. 234901, 2017.

22. Singh, Amit K., Mahesh P. Abegaonkar, and Shiban K. Koul, "Designing of wide angle beam steerable high gain flat top beam antenna using graded index metasurface," *IEEE Transactions on Antennas and Propagation*, pp. 1–1, June 2019.

7

Microwave Metamaterial Absorbers

7.1 Introduction

Microwave absorbers are the materials that absorb the EM energy [1]. Absorbers are used in a wide range of applications to eliminate unwanted radiations in defense applications. EM interference and EM compatibility have several limitations in practical applications. The advancement in the design of microwave absorbers using metamaterial is the solution to those limitations. EM metamaterials have a strong potential to control EM response of many microwave materials and devices by tailoring EM properties of material, such as permittivity and permeability, by designing proper UCs. Research focusing on metamaterials has generated devices such as super lens and invisibility cloaks for Electromagnetic Interference/ Electromagnetic compatibility (EMI/EMC) reduction, radar applications, anechoic chamber and medical applications. Metamaterials are used widely for antenna applications to enhance radiation characteristics of an antenna, for beam forming and beam steering. Metamaterial absorbers have attracted great attention due to their wide range of applications and easy fabrication. As metamaterial absorbers are built on large periodic arrays, their absorption responses depend on the incident as well as on the polarization angles of the incident EM waves [2]. However, practical applications require such kind of absorbing structures whose performance will remain constant over any incident and/or polarization angle. Therefore, microwave absorbers having angular and polarization angle stability characteristics are urgently required in the research fraternity. Narrow-band metamaterial absorber structures offer single or multiple absorption peaks, whose resonances are very sharp, and offer limited bandwidths. These structures are primarily used in different types of sensors, imaging instruments, bolometers and other related applications where sharp resonances are critical.

7.2 Working Principle

The EM waves impinging on the surface of an absorber induce surface currents on the surface of the UC. The direction of flow of surface current on UC and ground plane is found to be anti-parallel, resulting in circulating current loop formations and hence strong magnetic resonance occurs. Due to E-field concentrations on UC, electric resonance occurs. The common E- and H-resonance regions cause maximum EM absorption. A standard metal-backed metamaterial absorber is shown in Figure 7.1(a) with the equivalent circuit model in Figure 7.1(b).

In the equivalent circuit "Z_d" is impedance offered by dielectric of thickness "$t = 0.50$ mm".

The dielectric impedance "Z_d" is given by Equation 7.1.

$$Z_d = \frac{Z_0}{\sqrt{\varepsilon_r}} \tan(\beta t) \tag{7.1}$$

Where Z_0 is the characteristic impedance, β is the propagation constant and t is the substrate thickness. In the equivalent circuit Z_{uc} is the impedance due to proposed UC design. The total input impedance from Figure 7.1(b) is given by Equation 7.2.

$$Z_{in} = Z_{uc} \| Z_d \tag{7.2}$$

$$Z_{in} = (Z_{uc}) \| \frac{Z_0}{\sqrt{\varepsilon_r}} \tan(\beta t) \tag{7.3}$$

The input reflection coefficient Γ_{in} is given by Equation 7.4.

$$\Gamma_{in} = \frac{Z_{in} - Z_0}{Z_{in} + Z_0} \tag{7.4}$$

When equivalent parallel circuit resonates, then the overall reactance of Z_{uc} and Z_d becomes equal in magnitude but opposite in phase; this results in pure real value of Z_{in}. When pure real Z_{in} matches with free space impedance Z_0 then input reflection Γ_{in} becomes minimum. This minimum reflection results in maximum absorption. The frequency-dependent absorption is defined in Equation 7.5.

$$A(\omega) = 1 - T(\omega) - R(\omega) \tag{7.5}$$

Here $T(\omega) = |S_{21}(\omega)|^2$ is the transmission coefficient and $R(\omega) = |S_{11}(\omega)|^2$ is the reflection coefficient. Since the UC is metal-backed, $T(\omega) = 0$, and $A(\omega)$ reduces to Equation 7.6.

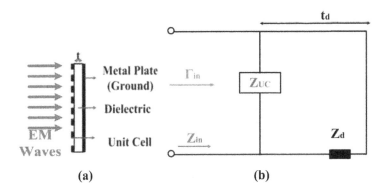

FIGURE 7.1
(a) Block diagram of metamaterial absorber and (b) equivalent circuit model of metamaterial absorber.

$$A(\omega) = 1 - R(\omega) \tag{7.6}$$

If $T(\omega_0) = R(\omega_0) = 0$, at some resonant frequency ω_0, then this will lead to maximum absorption.

7.3 Experimental Setup

To measure the performance of the absorber, a standard free space absorber measurement setup is used. A portable VNA (Anritsu MS2028C) and two wideband horn antennas as transmitter and receiver are used. The horn antennas used are separated by pyramidal foam absorber to reduce EM coupling at low frequency. The SUT is placed in the far-field of antenna to realize normal incidence on SUT. The setup is placed inside an anechoic chamber.

The measurement setup used is shown in Figure 7.2(a). The background noise of the anechoic chamber is measured first by recording reflection coefficient by VNA in the absence of any sample. The measured noise level is found to be below −65 dB over the entire desired frequency range as shown in Figure 7.2(b). The reflection coefficient of a metal plate of same dimension as that of the absorber is measured first. The metal plate is replaced with fabricated prototype absorber sample surface and reflection coefficient is measured. The results obtained are plotted in Figure 7.2(b). The actual reflection coefficient of fabricated sample absorber is the difference between reflection coefficients of metal plate and fabricated prototype absorber.

The standard free space absorber measurement setup is used to measure the absorption, and lock -in IR thermography is used to measure the temperature change on the absorber surface [3]. The metamaterial structures

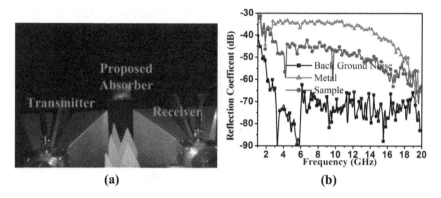

(a)   **(b)**

FIGURE 7.2
(a) Absorber measurement setup and (b) measured results.

are used to design multiband, ultrathin, polarization-insensitive microwave absorbers with narrow-band applications, and wideband absorption characteristics are obtained using passive loading as described in [1–16].

7.4 Penta-Band Polarization-Insensitive Metamaterial Absorber

In this section, a low-profile polarization-insensitive penta-band planar metamaterial microwave absorber for C-, X-, Ku- and Ka-bands is proposed. The absorber is extensively characterized in an anechoic chamber [17]. The simulated and measured results are found to be in good agreement.

7.4.1 Unit Cell Geometry and Simulated Results

The proposed absorber is designed on top of a 1.5-mm-thick FR-4 substrate with continuous metal backing. The metamaterial UC used is a compact low-profile double annular slot ring resonator (DASR). The design steps of DASR UC are explained in Chapter 2 as shown in Figure 2.10. DASR UC is excited by normal incidence EM waves on the top surface. The DASR UC with metal backing is simulated using periodic boundary condition using CST microwave studio. Simulated reflection coefficient with simulated absorptivity is shown in Figure 7.3. Normally incident EM waves on the surface of the UC generates resonance due to their equivalent inductive and capacitive behaviour. Resonance at 5.80 GHz, 10.84 GHz, 14.18 GHz, 17.41 GHz and 36.6 GHz with reflection coefficient dip of −22.98 dB, −29.41 dB, −18.89 dB, −14.87 dB and −30.12 dB, respectively, indicates absorption peak of 97.75 per cent, 97.40 per cent, 91.20 per cent, 95.96 per cent and 98.78 per cent, respectively, are observed.

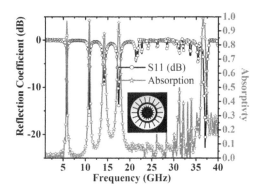

FIGURE 7.3
Simulated reflection coefficient and absorptivity.

Source: Copyright/used with permission of/courtesy of IEEE.

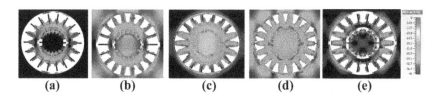

FIGURE 7.4
Simulated surface current distribution (a) 5.80 GHz; (b) 10.84 GHz; (c) 14.18 GHz; (d) 17.41 GHz and (e) 36.6 GHz.

Source: Copyright/used with permission of/courtesy of IEEE.

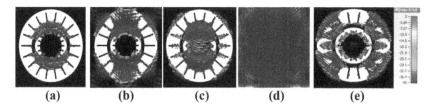

FIGURE 7.5
Simulated H-field distribution (a) 5.80 GHz; (b) 10.84 GHz; (c) 14.18 GHz; (d) 17.41 GHz and (e) 36.6 GHz.

Source: Copyright/used with permission of/courtesy of IEEE.

To get insight into the absorption phenomenon of the proposed DASR UC absorber, simulated surface current distribution, H-field distribution and E-field distribution are plotted in Figure 7.4, Figure 7.5 and Figure 7.6, respectively, for all the resonant bands. It is observed that absorption at 5.80 GHz is due to outer metallic ring thickness, at 10.84 GHz it is due to metallic serrations on inner ring, at 14.18 GHz it is due to inner ring, at 17.41 GHz it is due to

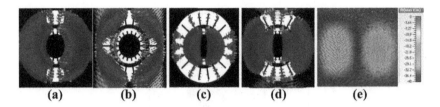

FIGURE 7.6
Simulated E-field distribution (a) 5.80 GHz; (b) 10.84 GHz; (c) 14.18 GHz; (d) 17.41 GHz and
(e) 36.6 GHz.

Source: Copyright/used with permission of/courtesy of IEEE.

metallic serrations on outer ring and at 36.51 GHz it is due to the gap between
metallic serrations on outer ring.

7.4.2 Measured Results

A 16 × 16 array of DASR metamaterial UC is printed on metal-backed FR-4
substrate with permittivity 4.3 and thickness of 1.5 mm. The fabricated
absorber is shown in Figure 7.7. The thickness of absorber is only $\dfrac{\lambda}{34.5}$, where

λ is the wavelength at first resonant frequency 5.80 GHz. The measurement setup, as discussed in Section 7.2, is used to measure the absorption
response. The fabricated absorber is tested for absorption response in an
anechoic chamber in two steps, first, using two similar standard gain wideband horn antennas, operating from 1 GHz to 20 GHz, and VNA and then
using Ka-band standard horn antenna next.

The measured reflection coefficient is compared with simulated reflection coefficient in Figure 7.8. Both results are found to be in good agreement.
The polarization of both horns is changed, and reflection from the proposed
absorber due to change in polarization angle is measured. The reflection
characteristics due to various angles of incidence of TE and TM-polarized
EM waves are measured and plotted in Figure 7.9 and Figure 7.10, respectively. The proposed absorber is found to be polarization-insensitive for
angular stability up to 45°.

The absorber is also measured for oblique incidence of EM waves and found
to be insensitive for up to 30° of oblique incidence as shown in Figure 7.11.

The fabricated absorber is also measured for its Ka-band absorption
response from 26 GHz to 40 GHz in an anechoic chamber. Two standard
Ka-band horn antennas separated by a pyramidal absorber and connected
with PNA E8364 VNA are used to record reflection as shown in Figure 7.12.
The background noise due to the anechoic chamber is recorded by measuring reflection in the absence of any sample and then reflection due to metal
plate is recorded and considered as reference for calibration. The reflection
due to FR-4 substrate without metal in both sides is recorded and uniform

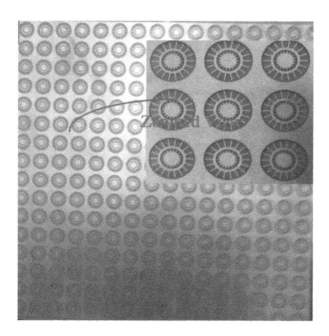

FIGURE 7.7
Fabricated penta-band absorber.

Source: Copyright/used with permission of/courtesy of IEEE.

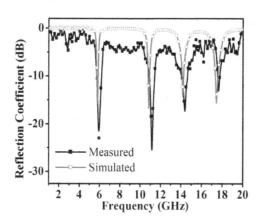

FIGURE 7.8
Simulated and measured reflection coefficients from 1 GHz to 20 GHz.

Source: Copyright/used with permission of/courtesy of IEEE.

absorption of about 3.5 dB is observed as shown in Figure 7.13. Then reflection due to fabricated absorber is recorded and all reflections are shown in Figure 7.13. The normalized reflection with respect to metal plate is actual reflection due to proposed metamaterial absorber.

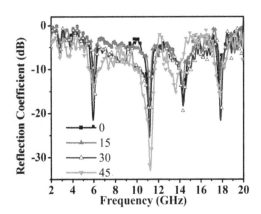

FIGURE 7.9
Measured reflection coefficient for various angles of incidence of TE-polarized EM waves from 1 GHz to 20 GHz.

Source: Copyright/used with permission of/courtesy of IEEE.

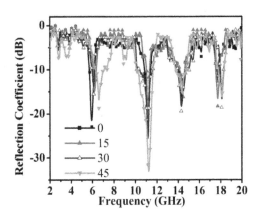

FIGURE 7.10
Measured reflection coefficient for various angles of incidence of TM-polarized EM waves from 1 GHz to 20 GHz.

Source: Copyright/used with permission of/courtesy of IEEE.

Measured reflection coefficient and simulated reflection coefficient of proposed absorber for Ka-band are compared is Figure 7.14 and found to be in good agreement. The measured reflection characteristics due to various angles of incidence of TE- and TM-polarized EM waves are recorded and shown in Figure 7.15 and Figure 7.16, respectively.

The measured reflection due to oblique incidence of EM waves is recorded and plotted in Figure 7.17. The designed absorber is insensitive to the oblique incidence of EM waves up to an angular stability of 45°.

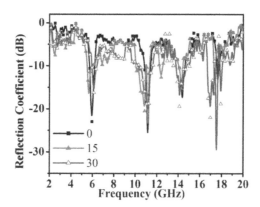

FIGURE 7.11
Measured reflection coefficient for various oblique incidence of EM waves from 1 GHz to 20 GHz.

Source: Copyright/used with permission of/courtesy of IEEE.

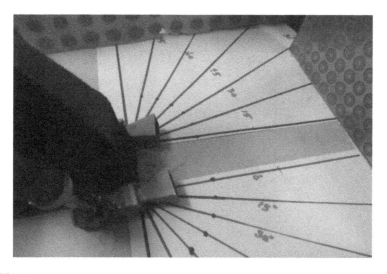

FIGURE 7.12
Measurement setup for Ka-band from 26 GHz to 40 GHz.

Source: Copyright/used with permission of/courtesy of IEEE.

A penta-band polarization-insensitive metamaterial planar microwave absorber is successfully demonstrated. The measured bandwidth with more than 90 per cent absorption is 0.15 GHz, 0.21 GHz, 0.19 GHz, 0.35 GHz and .891 GHz at absorption resonant frequencies of 5.91 GHz, 11.14 GHz, 14.28 GHz, 17.61 GHz and 36.8 GHz, respectively. The proposed absorber is polarization-insensitive to TE- and TM-polarized incidence of EM waves

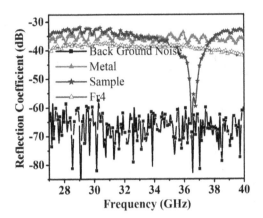

FIGURE 7.13
Actual measured reflection coefficient from 26 GHz to 40 GHz.

Source: Copyright/used with permission of/courtesy of IEEE.

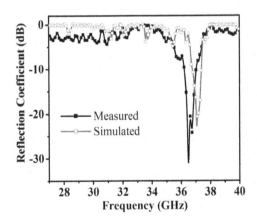

FIGURE 7.14
Simulated and measured reflection coefficient of the proposed absorber from 26 GHz to
40 GHz.

Source: Copyright/used with permission of/courtesy of IEEE.

with wide measured angular stability up to 45° for the first, second, third
and fourth bands and up to 30° for the fifth band. The absorber is insensitive to the oblique incidence of EM waves up to the measured angular stability of 30° for the first, second, third and fourth bands and up to 45° for
the fifth band [17].

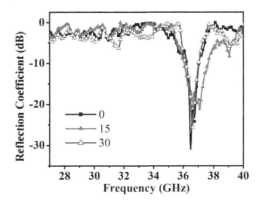

FIGURE 7.15
Measured reflection coefficient for various angles of incidence of TE-polarized EM waves from 26 GHz to 40 GHz.

Source: Copyright/used with permission of/courtesy of IEEE.

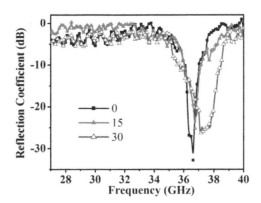

FIGURE. 7.16
Measured reflection coefficient at various angles of incidence of TM-polarized EM waves from 26 GHz to 40 GHz.

Source: Copyright/used with permission of/courtesy of IEEE.

7.5 Triple-Band Polarization-Insensitive Ultrathin Metamaterial Absorber for S-, C- and X-Band Applications

This section describes the design of a triple-band ultrathin polarization-insensitive metamaterial absorber for S-, C- and X-band applications [18]. The absorber is characterized in an anechoic chamber and absorption phenomenon is demonstrated.

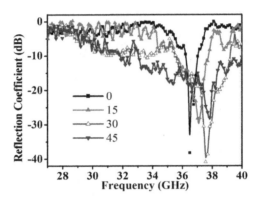

FIGURE 7.17
Measured reflection coefficient for various oblique incidence of EM waves from 26 GHz to 40 GHz.

Source: Copyright/used with permission of/courtesy of IEEE.

7.5.1 Unit Cell and Simulated Results

The proposed metamaterial absorber is made up of a modified triple CsRR as UC. A triple CRR is proposed first. The perimeter of rings is selected as one-half of the wavelength at optimum resonant frequency. Radial slots are introduced on the metallic ring to down convert the frequency. It is observed that a compactness of 22.50 per cent is achieved due to outer MCRR to have absorption centre frequency at 2.9 GHz as compared to conventional CRR. The proposed modified triple circular slot ring resonator (MTCsRR) UC geometry with detailed dimensions is shown in Figure 7.18(a). The MTCsRR UC is simulated using CST microwave studio with perfect electric boundary along Y-axis, perfect magnetic boundary along X-axis and a WG port along the Z-axis as shown Figure 7.18(b). A simulated triple-band resonance with centre frequency of 2.90 GHz, 4.20 GHz and 9.30 GHz with −10-dB reflection bandwidth of 0.27 GHz, 0.24 GHz and 0.28 GHz, respectively, is observed, which indicates that more than 90 per cent absorption is obtained as shown in Figure 7.19.

To validate the cause of resonance at these frequencies, the simulated H-field and E-field distributions on MTCsRR are studied and plotted in Figure 7.20 and Figure 7.21, respectively. The normal incident plane wave (along the Z-axis) having E-field along the X-axis induces capacitive effect and H-field along the Y-axis induces inductive effect, resulting in resonance. The maximum concentration of E- and H-fields at first resonant frequency of 2.9 GHz is found to be on outer MCRR, at second resonant frequency of 4.2 GHz on middle MCRR and at third resonant frequency of 9.2 GHz on inner MCRR. The circulating H-field induces different surface currents on the MTCsRR at different frequencies. The surface current distribution at all three resonant frequencies is plotted in Figure 7.22. It is observed that the surface currents on MTCsRR and ground plane at resonant frequencies are anti-parallel to each other, forming a loop and resulting in absorption of respective frequency.

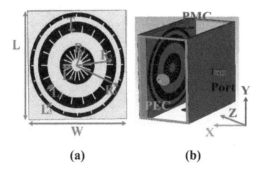

(a)   **(b)**

FIGURE 7.18

(a) Geometry of proposed MTCsRR UC and (b) the boundary conditions to simulate MTCsRR UC. The dimensions are $R_1 = 2.00$, $R_2 = 5.00$, $R_3 = 6.50$, $L_1 = 1.45$, $L_2 = 1.10$, $L_3 = 0.40$, $T = 0.20$, $L = 14.00$ and $W = 14.00$ (all dimensions are in millimetres).

Source: Copyright/used with permission of/courtesy of EMW.

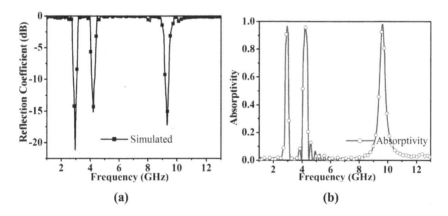

(a)   **(b)**

FIGURE 7.19

(a) Simulated reflection coefficient and (b) simulated absorptivity.

Source: Copyright/used with permission of/courtesy of EMW.

The E-field, H-field and surface current distribution predict that the first resonant band is generated due to outer MCRR, second resonant band is generated due to middle MCRR and third resonant band is generated due to inner MCRR. The change in dimensions of MCRR will result in a change in the respective resonant frequency. The mean radius of outer MCRR "R_1", middle MCRR "R_2" and inner MCRR "R_3" is varied, and the corresponding changes in reflection coefficients are plotted in Figure 7.23(a), Figure 7.23(b) and Figure 7.23(c), respectively. It is observed that increasing the mean radius of MCRR increases the ring perimeter, causing an increase in effective inductance and capacitance, thereby resulting in a decrease in the resonant frequency. Similarly, decreasing the mean radius of MCRR increases resonant

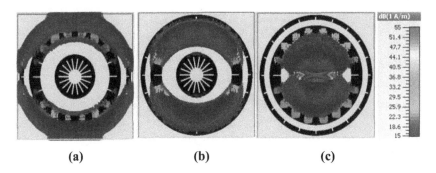

(a)   (b)   (c)

FIGURE 7.20
Simulated H-field distribution (a) 2.9 GHz; (b) 4.2 GHz and (c) 9.2 GHz.

Source: Copyright/used with permission of/courtesy of EMW.

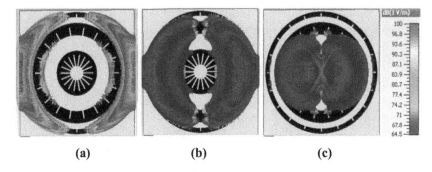

(a)   (b)   (c)

FIGURE 7.21
Simulated E-field distribution (a) 2.9 GHz (b) 4.2 GHz and (c) 9.2 GHz.

Source: Copyright/used with permission of/courtesy of EMW.

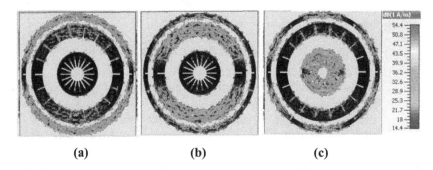

(a)   (b)   (c)

FIGURE 7.22
Simulated surface current distribution (a) 2.9 GHz; (b) 4.2 GHz and (c) 9.2 GHz.

Source: Copyright/used with permission of/courtesy of EMW.

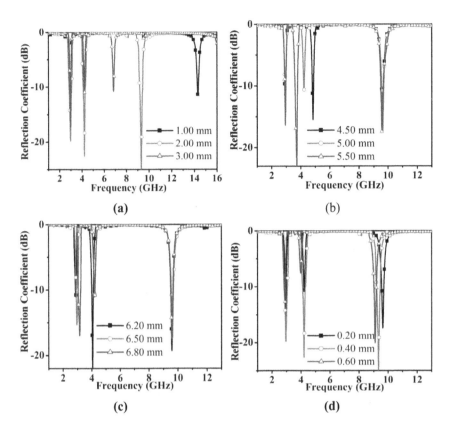

FIGURE 7.23
Simulated reflection coefficient with variation in mean radius of (a) inner ring resonator "R_1"; (b) middle ring resonator "R_2"; (c) outer ring resonator "R_3" and (d) simulated reflection coefficient with variation in slot thickness.

Source: Copyright/used with permission of/courtesy of EMW.

frequency. The slots on the metallic ring act as a capacitive load due to in-plane E-field on MCRR. The effect of dimensional change of slot is studied and plotted in Figure 7.23(d). It is observed that the change in slot dimensions changes all the resonant frequencies, thereby predicting that all the resonant frequencies depend on the slot dimensions. It is observed by the parametric variations in Figure 7.23 that all the three absorption bands can be controlled independently.

7.5.2 Measured Results

The prototype of ultrathin triple-band metamaterial absorber is printed using photolithography on FR-4 substrate having permittivity of 4.3, dielectric thickness of 0.762 mm and copper thickness of 0.035 mm. The fabricated prototype of metamaterial absorber is shown in Figure 7.24. The absorption response of the absorber is measured in an anechoic chamber using the measurement setup shown in Figure 7.2. The back-ground noise of the anechoic chamber is measured first. The background noise is observed to be less than −60 dB. The reflection due to metal plate and absorber is recorded next as shown in Figure 7.25. The simulated and measured reflection characteristics of the proposed absorber are plotted in Figure 7.26. The results are found to be in good agreement. Measured− 10-dB reflection band of 0.40 GHz at first absorption frequency of 2.9 GHz, 0.45 GHz at second absorption frequency of 4.18 GHz and 0.47 GHz at third absorption frequency of 9.25 GHz is observed predicting more than 90 per cent absorption.

The proposed absorber is studied for various angles of incidence of EM waves with TE- and TM-polarization. The angle between both double-ridge

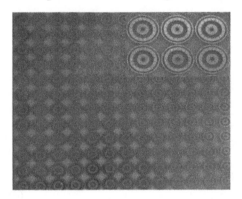

FIGURE 7.24
Fabricated prototype absorber.

Source: Copyright/used with permission of/courtesy of EMW.

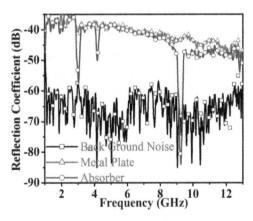

FIGURE 7.25
The measured reflection coefficient of anechoic chamber, metal plate and absorber sample.

Source: Copyright/used with permission of/courtesy of EMW.

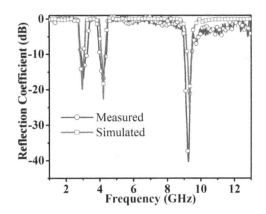

FIGURE 7.26
Measured and simulated reflection coefficients due to absorber prototype.

Source: Copyright/used with permission of/courtesy of EMW.

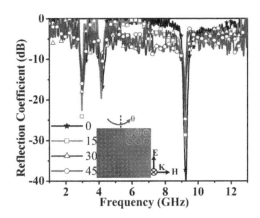

FIGURE 7.27
Measured reflection coefficient due to various TE-polarized EM waves.

Source: Copyright/used with permission of/courtesy of EMW.

horn antennas is changed in its TE and TM planes to generate EM waves with various angles of incidence of TE- and TM-polarized EM waves. The reflection characteristics of prototype absorber due to various angles of incidence of TE- and TM-polarized EM waves are measured and plotted in Figure 7.27 and Figure 7.28, respectively. It is observed that the proposed absorber is insensitive to various angles of incidence of TE- and TM-polarized EM waves with wide angular stability up to 45°. The absorber is studied for various oblique angles of incidences of EM waves in the elevation plane. The measured reflection characteristics due to various oblique angles of incidence of EM waves are plotted in Figure 7.29. It is observed that the absorber is insensitive to various oblique angles of incidence of EM waves up to 45°.

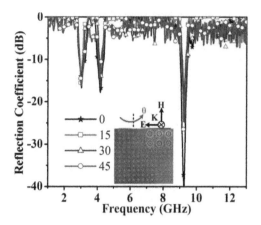

FIGURE 7.28
Measured reflection coefficient due to various TM-polarized EM waves.

Source: Copyright/used with permission of/courtesy of EMW.

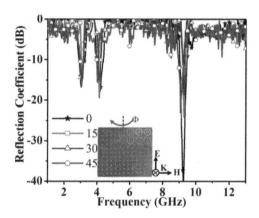

FIGURE 7.29
Measured reflection coefficient due to various oblique incidence of EM waves.

Source: Copyright/used with permission of/courtesy of EMW.

A triple-band polarization-insensitive ultrathin metamaterial absorber for S-, C- and X-band applications is successfully demonstrated. The proposed absorber is ultrathin having thickness of $\lambda_0/135.66$ at lowest absorption centre frequency of 2.90 GHz. The absorption bandwidths of 0.40 GHz, 0.45 GHz and 0.47 GHz with more than 90 per cent absorption at absorption centre frequencies of 2.90 GHz, 4.18 GHz and 9.25 GHz having absorption peak of 97 per cent, 96.45 per cent and 98.20 per cent, respectively, are observed. The proposed absorber is found to be insensitive to the oblique incidence of EM waves, TE- and TM-polarized various angles of incidence of the EM waves.

7.6 Conformal Ultrathin Polarization-Insensitive Double-Band Metamaterial Absorber

This section describes the design of a dual-band conformal ultrathin EM absorber for C- and X-band applications [18]. The fabricated prototype is extensively measured and absorption phenomenon is verified.

7.6.1 Unit Cell Geometry and Simulation Results

The proposed UC consists of two CRRs which are modified to make the absorber compact. Initially, metallic rings on a grounded substrate are designed. The circumference of the rings is approximately equal to one guide-wavelength at respective frequencies of 4.75 GHz and 14.45 GHz. The closed metallic ring produces self-inductance. The magnetic flux incident on the metallic ring generates rotating currents which induce their own magnetic flux. To support larger wavelength to propagate or get absorbed and to achieve compactness, the closed metallic rings can be loaded with additional capacitors. The loading capacitor is designed by modifying the ring with periodic slots as seen in Figure 7.30(a). The slots provide additional capacitance, resulting in *LC* resonant circuit. The circular ring having equivalent self-inductance and capacitance due to periodic slots acts like an MCRR. The effective capacitance of single slot on MCRR ring can be given by [19].

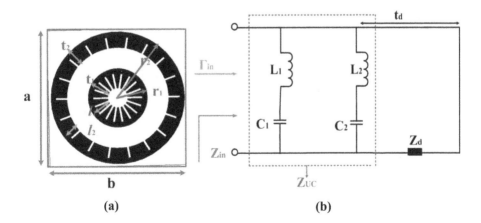

FIGURE 7.30

(a) MDCRR UC and (b) the equivalent circuit model of MDCRR UC. Dimensions are: $a = 14.0$, $b = 14.0, r_1 = 3.0, r_2 = 6.5, L_1 = 1.51, L_2 = 1.10, t_1 = 2.00$ and $t_2 = 1.60$ (all dimensions are in millimetres).

Source: Copyright/used with permission of/courtesy of IEEE.

$$C = \varepsilon_0 \varepsilon_{\text{eff}} \frac{2l_i}{\pi} \ln\left(\frac{1}{\sin \frac{\pi d}{2l_i}} \right) \tag{7.7}$$

Here, $d = 2\pi r_i / N$, and N = number of strips.

For the MCRR, 18 slots are in series and hence total equivalent capacitance can be calculated. The number of slots and dimension of slots are optimized to get the desired resonance frequency using CST microwave studio. The self-inductance of a strip of length l and width $t_i(i = 1,2,3)$ is given by Equation 7.8. The effective inductance of ring circumference can be obtained by Equation 7.9 as in [19].

$$L_i = 2 \times 10^{-4} l \left[\ln\left(\frac{l}{t_i + t} \right) + 1.193 + 0.2235 \frac{t_i + t}{l} \right] \tag{7.8}$$

$$L = \mu_0 \mu_{\text{eff}} \frac{r_i}{2\pi} \ln\left(\frac{1}{\sin \frac{\pi \omega}{2r_i}} \right) \tag{7.9}$$

The calculated circuit parameters are L_1 = 46.22 nH, L_2 = 93.50 nH, C_1 = 4.97 fF and C_2 = 3.62 fF.

A compactness of 11.80 per cent and 32.50 per cent is achieved with resonating frequencies of 4.19 GHz and 9.75 GHz, respectively. The proposed MDCRR UC is complementary to double annular slot ring resonator (DASR) presented in Chapter 2, Section 2.5.1. The black-coloured section represents the metal. The MDCRR UC is printed on GML 1000 substrate having permittivity 3.2 and thickness of 0.502 mm on one side with another side of substrate acting as continuous metal ground. The MDCRR UC is simulated using the same boundary conditions as discussed in Section 7.4.2. The equivalent circuit diagram of the MDCRR UC is realized and is shown in Figure 7.30(b).

In the equivalent circuit, Z_{uc} is impedance due to proposed UC. "L_1" is the effective inductance due to inner MCRR and "C_1" is effective capacitance due to slots on inner MCRR. "L_2" is the effective inductance due to outer MCRR and "C_2" is effective capacitance due to slots on outer MCRR and gaps between the adjacent UCs as shown in Figure 7.30(b).

The total input impedance from Figure 7.30(b) is given by Equation 7.10.

$$Z_{\text{in}} = Z_{\text{uc}} \| Z_{\text{d}} \tag{7.10}$$

$$Z_{in} = (j\omega L_1 + \frac{1}{j\omega C_1}) \ | \ | \ (j\omega L_2 + \frac{1}{j\omega C_2}) \ | \ | \ \frac{Z_0}{\sqrt{\varepsilon}_r} \tan(\beta t) \dots \dots \text{for}$$

MDCRR UC.

The proposed MDCRR UC is simulated with periodic boundary condition under normal incidence of EM waves using CST microwave studio software. The EM-simulated reflection coefficient obtained using CST microwave studio software and circuit-simulated reflection coefficient obtained using ADS software (using circuit parameters given earlier) are compared in Figure 7.31(a). The simulated absorption is plotted in Figure 7.31(b).

Dual-band resonance is observed due to MDCRR UC with resonating frequencies at 4.19 GHz and 9.75 GHz with reflection coefficient of −19.16 dB and −24.93 dB and absorption of 97.21 per cent and 98.85 per cent, respectively. The circuit-simulated reflection coefficient is deviated from EM-simulated reflection coefficient as the parasitics were neglected in the equivalent circuit realizations.

To get insight into absorption phenomenon of the proposed structure, E-field and H-field distributions at both absorption bands 4.19 GHz and 9.75 GHz are plotted in Figure 7.32 and Figure 7.33, respectively. It is observed that resonance at 4.15 GHz is due to E- and H-field concentrations on outer MCRR. The resonance at 9.75 GHz is due to E- and H-field concentrations on inner MCRR. The surface current distribution due to normal incidence of EM waves is studied in Figure 7.34(a) and (b). The surface current distribution is concentrated on outer MCRR at 4.19 GHz and on inner MCRR at 9.75 GHz. The field and surface current distributions predict that the lower

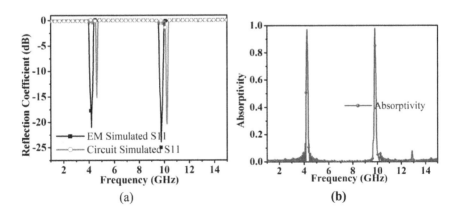

FIGURE 7.31
(a) The EM- and circuit-simulated reflection coefficients of MDCRR UC and (b) the simulated absorptivity.

Source: Copyright/used with permission of/courtesy of IEEE.

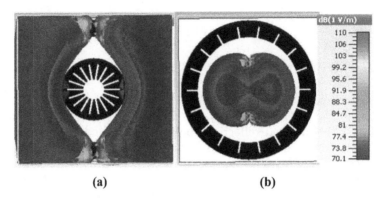

FIGURE 7.32
Simulated E-field distribution on MDCRR (a) 4.19 GHz and (b) 9.75 GHz.

Source: Copyright/used with permission of/courtesy of IEEE.

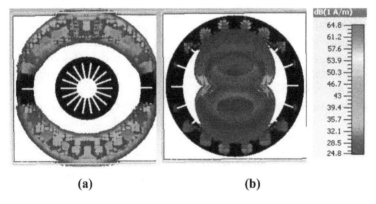

FIGURE 7.33
Simulated H-field distribution on MDCRR (a) 4.19 GHz and (b) 9.75 GHz.

Source: Copyright/used with permission of/courtesy of IEEE.

band of 4.19 GHz is generated due to outer MCRR and higher band of 9.75 GHz is generated due to inner MCRR.

The parametric variation of radius of outer MCRR "r_2" and inner MCRR "r_1" is carried out and the results are plotted in Figure 7.35(a) and Figure 7.35 (b), respectively. Increase in mean radius increases equivalent inductance and capacitance, resulting in a decrease in the resonant frequency. The parametric variations predict that both absorption bands can be controlled independently.

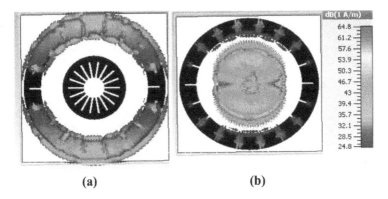

FIGURE 7.34
Simulated surface current distribution on MDCRR (a) 4.19 GHz and (b) 9.75 GHz.

Source: Copyright/used with permission of/courtesy of IEEE.

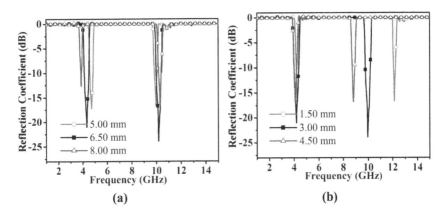

FIGURE 7.35
Simulated reflection coefficient (a) variation of radius of outer MCRR "r_2" and (b) variation of radius of inner MCRR "r_1".

Source: Copyright/used with permission of/courtesy of IEEE.

7.6.2 Measured Results

The prototypes of conformal double-band absorber are fabricated and tested in an anechoic chamber. A 16 × 16 array of the proposed MDCRR UC is fabricated on GML 1000 substrate having thickness of 0.50 mm and permittivity of 3.2. The fabricated prototypes of dual-band absorber with flat and curved surfaces are shown in Figure 7.3(a) and (b), respectively. The fabricated absorber is a periodic arrangement of proposed MDCRR UCs.

To measure the performance of the proposed absorber, the measurement setup as discussed in Section 7.2 is used. The background noise of

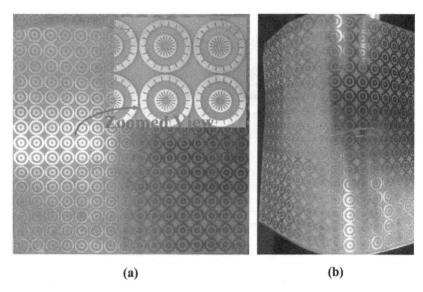

(a)   **(b)**

FIGURE 7.36
Fabricated dual-band flat and curved surfaces of proposed absorber. (a) Flat surface of proposed absorber and (b) curved surface of proposed absorber.

Source: Copyright/used with permission of/courtesy of IEEE.

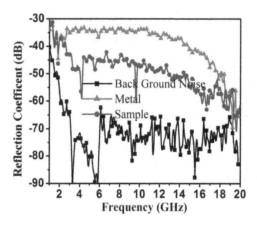

FIGURE 7.37
Measured background noise and reflection coefficient due to metal plate and fabricated double-band prototype absorber.

Source: Copyright/used with permission of/courtesy of IEEE.

anechoic chamber is measured by recording reflection coefficient by VNA in the absence of any sample. The measured noise level is below −65 dB over the entire desired frequency range as shown in Figure 7.37. The reflection coefficient of a metal plate of the same dimension as that of the absorber is

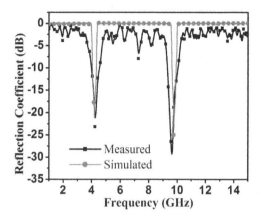

FIGURE 7.38
Simulated and measured reflection coefficient of proposed double-band absorber.

Source: Copyright/used with permission of/courtesy of IEEE.

measured first. The metal plate is replaced with a fabricated flat double-band prototype absorber surface and reflection coefficient is again measured. The results obtained are plotted in Figure 7.37.

The measured and simulated results of double-band metamaterial absorber are plotted in Figure 7.38. Good agreement between simulated and measured results is observed. The measured –10-dB impedance bandwidth of double-band metamaterial absorber is 0.67 GHz and 0.86 GHz at centre frequencies of 4.19 GHz and 9.75 GHz, respectively, with more than 90 per cent absorption. The reflection (equivalently the absorption) from the absorber for TM and TE field polarization when angle between the transmitting and receiving antennas is changed is studied (this is equivalent to changing the angle of incidence in azimuth plane). The measured reflection coefficients of double-band absorber for TM-polarized wave and TE-polarized waves with varying angle between the transmitter and receiver are shown in Figure 7.39(a) and (b), respectively. It is seen that the measured results are in good agreement. From the figures, it can be inferred that the proposed absorber has a wide angular stability up to 45°.

The proposed triple-band absorber structure is studied for various oblique angles of incidence in the elevation plane. The measured results are plotted in Figure 7.40. The results are in good agreement. Hence, the proposed absorber is insensitive to polarization and angle of incidence of the EM waves. The polarization-insensitivity is attributed to four-fold symmetry of MTCRR UC. Similar results are obtained for MDCRR absorber.

To verify the conformal characteristics of the proposed absorber, the absorber is pasted on a curved surface and reflection coefficient is recorded under normal incidence. The same process is repeated by pasting a metal

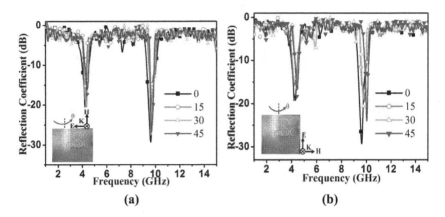

FIGURE 7.39
Measured reflection coefficient due to (a) TM-polarized angle of incidence of EM waves and
(b) TE-polarized angle of incidence of EM waves.

Source: Copyright/used with permission of/courtesy of IEEE.

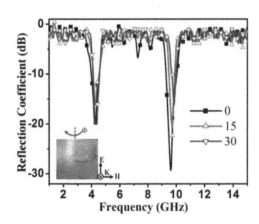

FIGURE 7.40
Measured reflection coefficient due to various oblique angles of incidence of EM waves.

Source: Copyright/used with permission of/courtesy of IEEE.

sheet on the same curved surface. The fabricated absorber is pasted on a
curved surface with curvature angle of about 120°. The actual reflection coefficient of curved surface absorber is the difference between the two reflection coefficients.

The reflection coefficients of the proposed absorber for flat surface and
curved surface are compared with the simulated flat surface in Figure 7.41
for the proposed double-band conformal absorber. The compared results are
in good agreement. Hence, the proposed double-band absorber surface is
conformal.

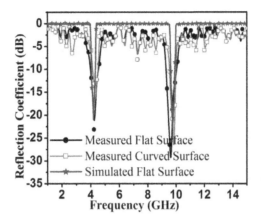

FIGURE 7.41
Measured reflection coefficient of proposed fabricated prototype double-band absorber with flat surface and curved surface compared with simulated flat absorber.

Source: Copyright/used with permission of/courtesy of IEEE.

The design of a double-band conformal ultrathin polarization-insensitive metamaterial absorber is successfully demonstrated. The absorption peaks of 97.21 per cent and 98.85 per centat 4.19 GHz and 9.75 GHz, respectively, are observed. The absorption bandwidths at these frequencies are 0.67 GHz and 0.86 GHz, respectively, with more than 90 per cent absorption. The thickness of the proposed absorber surface is $\lambda_0 / 142.65$, where λ_0 is the free space wavelength at first centre frequency of absorption (4.19 GHz), making the absorber ultrathin. The conformal behaviour of the proposed absorber makes it suitable for a lot of practical applications.

7.7 Triple-Band Polarization-Insensitive Ultrathin Conformal Metamaterial Absorber with Wide Angular Stability

In this section, triple-band EM absorber for C- and X-band applications is presented. The absorber is experimentally characterized to be conformal and polarization-insensitive. The proposed absorber is ultrathin with a thickness of $\lambda_0 / 142.65$ at lowest frequency of absorption. The temperature profile of the proposed absorber is measured using lock-in IR thermography.

7.7.1 Design and Working Principle

In this section, the design of the proposed novel modified triple circular ring resonator (MTCRR) metamaterial UC with its absorption mechanism is discussed. Initially, metallic rings on a grounded substrate are designed. The circumference of the rings is approximately equal to one guide-wavelength at respective frequencies of 4.19 GHz, 6.64 GHz and 9.94 GHz. The closed metallic rings will produce self-inductance. The magnetic flux incident on the metallic ring generates rotating currents which induce their own magnetic flux. To support a larger wavelength to propagate or get absorbed and to achieve compactness, the closed metallic rings can be loaded with additional capacitors. The loading capacitor is designed by modifying the ring with periodic slots as seen in Figure 7.42(a). Compactness of 11.80 per cent, 26.35 per cent and 32.50 per cent is achieved with resonating frequencies 4.19 GHz, 6.64 GHz and 9.94 GHz, respectively. The slots provide additional capacitances, resulting in LC resonant circuit. The proposed MTCRR UC is modified complementary to double annular slot ring resonator (DASR). The geometrical detail of MTCRR UC is shown in Figure 7.42(a), where black colour represents the metal. The UC is printed on one side of substrate with another side of substrate acting as a continuous metal ground.

The equivalent circuit of the proposed metal-backed MTCRR UC is given in Figure 7.42(b). In the equivalent circuit "Z_d" is the impedance offered by dielectric of thickness "$t = 0.50$ mm" and Z_{uc} is the impedance due to proposed UC design. "L_1" is the effective inductance due to inner MCRR and "C_1" is effective capacitance due to slots on inner MCRR. "L_2" is the effective inductance due to middle MCRR and "C_2" is the effective capacitance due to outer MCRR and middle MCRR of MTCRR UC. "L_3" is the effective inductance due to outer MCRR and "C_3" is effective capacitance due to slots on outer MCRR and gaps between the adjacent UCs as shown in Figure 7.42. The values of effective inductances and capacitances are calculated using fundamental equations as reported in [1], [17]. The calculated circuit parameters are $L_1 = 46.22$ nH, $L_2 = 0.358$ nH, $L_3 = 93.50$ nH, $C_1 = 4.97$ fF, $C_2 = 2.30$ pF and $C_3 = 3.62$ fF. The proposed MTCRR UC is simulated with periodic boundary condition under normal incidence of EM waves in CST microwave studio software. The EM-simulated reflection coefficient obtained using CST microwave studio software and circuit-simulated reflection coefficient obtained using ADS software (using circuit parameters given earlier) are plotted in Figure 7.43(a). Triple-band resonance is observed with MTCRR UC at 4.19 GHz, 6.64 GHz and 9.94 GHz with reflection coefficients of –15.87 dB, –17.50 dB and –20.25 dB and absorption of 97.50 per cent, 96.50 per cent and 98.85 per cent, respectively, as shown in Figure 7.43(b). The circuit-simulated reflection coefficient is deviated from EM-simulated reflection coefficient as the parasitics are neglected in the equivalent circuit.

To get an insight into the generated resonance band, E-field distribution, H-field distribution and surface current distribution on the UCs are observed and studied. As MTCRR UC is a modified form of MDCRR UC, the surface current and field distributions on both UCs will remain similar

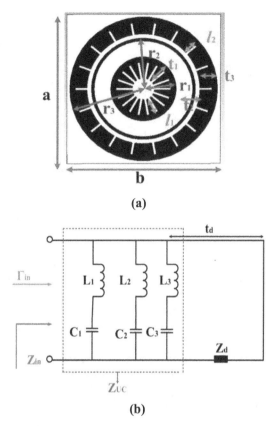

(a)

(b)

FIGURE 7.42

(a) MTCRR UC and (b) the equivalent circuit model of MTCRR UC. Dimensions are: $a = 14.0$, $b = 14.0$, $r_1 = 3.0$, $r_2 = 4.75$, $r_3 = 6.75$ $l_1 = 1.51$, $l_2 = 1.10$, $t = 0.50$, $t_1 = 2.0$, $t_2 = 0.25$, $t_3 = 1.6$ (all dimensions are in millimetres).

Source: Copyright/used with permission of/courtesy of IEEE.

at the corresponding resonating frequencies except at the modified region. Figure 7.44, Figure 7.45 and Figure 7.46 depict the E-field distribution, the H-field distribution and the surface current distribution, respectively.

It is seen that the fields as well as surface currents are concentrated on outer MCRR at 4.19 GHz, on middle MCRR at 6.64 GHz and on inner MCRR at 9.95 GHz.

The field and current distribution plots suggest that the resonance band at 4.19 GHz is generated due to outer MCRR, the resonance band at 6.64 GHz is generated due to middle MCRR and the resonance band at 9.95 GHz is generated due to inner MCRR. To verify the results mentioned earlier, parametric variation of radius of outer MCRR, middle MCRR and inner MCRR of MTCRR UC is carried out and the results are plotted in Figure 7.47(a), Figure 7.47(b) and Figure 7.47(c), respectively.

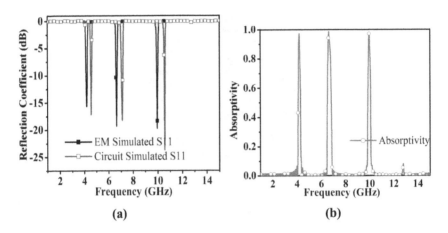

FIGURE 7.43
(a) EM- and circuit-simulated reflection coefficients of MTCRR UC and (b) simulated absorptivity.

Source: Copyright/used with permission of/courtesy of IEEE.

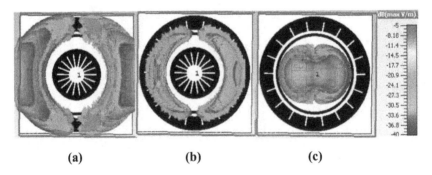

FIGURE 7.44
Electric field distribution at (a) 4.19 GHz; (b) 6.64 GHz and (c) 9.94 GHz.

Source: Copyright/used with permission of/courtesy of IEEE.

The parametric variations suggest that all the three absorption bands can be controlled independently.

The capacitance of slot depends on the slot dimension and number of slot. The number and dimensions of slots are optimized to get the desired frequency band. The decrease in the number of slots causes decrease in the effective capacitance and hence an increase of the resonant frequencies and vice versa as shown in Figure 7.47(d). The increase in area of slot (length and width) results in decrease in resonant frequency due to increase in effective capacitance and vice versa as shown in Figure 7.48(a) and (b). Fifty per cent change in slot width will cause change of 5.9 per cent,

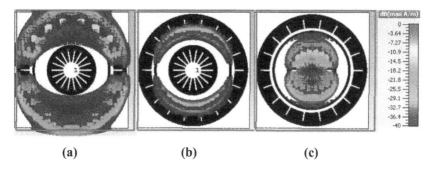

(a) **(b)** **(c)**

FIGURE 7.45
Magnetic field distribution at (a) 4.19 GHz; (b) 6.64 GHz and (c) 9.94 GHz.

Source: Copyright/used with permission of/courtesy of IEEE.

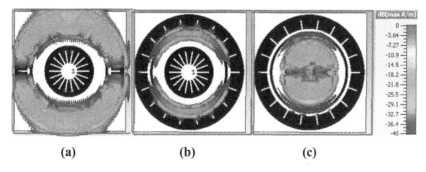

(a) **(b)** **(c)**

FIGURE 7.46
Surface current distribution at (a) 4.19 GHz; (b) 6.64 GHz and (c) 9.94 GHz.

Source: Copyright/used with permission of/courtesy of IEEE.

1.5 per cent and 1.9 per cent in the first, second and third resonant frequencies, respectively. Twenty per cent change in slot length will cause change of 4.75 per cent, 1.2 per cent and 3.5 per cent in first, second and third resonant frequencies, respectively. Hence, the proposed absorber can be considered as fabrication-tolerant.

7.7.2 Measured Results

The prototypes of conformal ultrathin triple band absorber are fabricated and tested in an anechoic chamber. A 16 × 16 array of the proposed MTCRR UC is fabricated on GML 1000 substrate with thickness of 0.50 mm and permittivity of 3.2. The fabricated absorber is a periodic arrangement of proposed MTCRR UCs. To measure the performance of the proposed absorber, the reflection coefficient due to prototype absorber is measured using the standard free space technique as discussed in Section 7.2. The measured

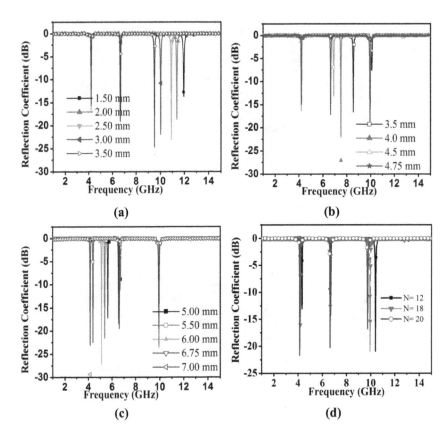

FIGURE 7.47
Reflection coefficient verses frequency plot at different (a) r_1 for MTCRR UC; (b) r_2 for MTCRR UC; (c) r_3 for MTCRR UC and (d) number of slots.

Source: Copyright/used with permission of/courtesy of IEEE.

and simulated reflection coefficients of the prototype absorber are compared in Figure 7.49. The simulated and measured results are found to be in good agreement. The measured −10-dB impedance bandwidth of the triple-band metamaterial absorber, with absorption more than 90 per cent, is 0.61 GHz around 4.19 GHz, 0.67 GHz around 6.64 GHz and 0.57 GHz around 9.95 GHz.

The reflection (equivalently the absorption) from the absorber for TE and TM field polarization is measured and plotted in Figure 7.50(a) and (b), respectively. It is seen that the measured and simulated results are in good agreement with wide angular stability up to 45°. The proposed triple-band absorber structure is studied for various oblique angles of incidence in elevation plane having measured results as shown in Figure 7.50(c). The measured results are found to be in good agreement with simulated

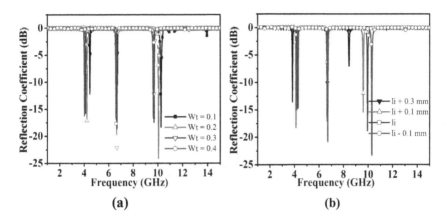

FIGURE 7.48
Reflection coefficient for different (a) slot width "W_t" variations and (b) slot length "L_i" variations (where i = 1, 2) for MTCRR UC.

Source: Copyright/used with permission of/courtesy of IEEE.

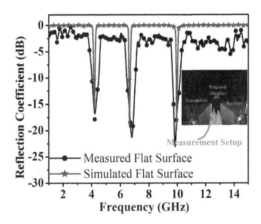

FIGURE 7.49
Simulated and measured reflection coefficients of MTCRR absorber.

Source: Copyright/used with permission of/courtesy of IEEE.

results. Hence, the proposed absorber is insensitive to polarization and angle of incidence of the EM waves. The polarization-insensitivity is attributed to four-fold symmetry of the MTCRR UC.

To verify the conformal characteristics of the proposed absorber, the absorber is pasted on a curved surface, and reflection coefficient is recorded under normal incidence. The same process is repeated by pasting a metal sheet on the same curved surface. The fabricated absorber is pasted on a curved surface with curvature angle of about 120°. The actual reflection

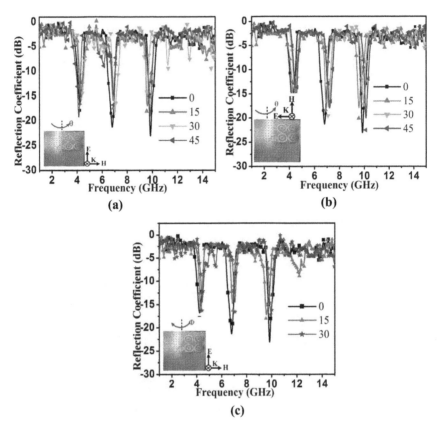

FIGURE 7.50
The measured reflection coefficient (equivalently the absorption) (a) TE-polarized wave; (b) TM-polarized wave and (c) oblique angle of incidence.

Source: Copyright/used with permission of/courtesy of IEEE.

coefficient of the curved-surface absorber is the difference between both reflection coefficients. The reflection coefficients of the proposed absorber for flat surface and curved surface are compared with simulated flat surface in Figure 7.51. The compared results are in good agreement. Hence, the proposed triple-band absorber surface is conformal.

It is expected that when the absorber absorbs the EM waves, the result should reflect an increase in its temperature on the absorber surface. To confirm the absorption properties of the proposed structure, its temperature profile is measured using lock-in IR thermography as reported in [19]. The prototype triple-band absorber is excited by a wideband double-ridge horn antenna acting as a microwave source, and temperature profile is plotted by thermal images of absorber screen as recorded by IR camera. The RF power is set to 10 dBm. The measurement setup is shown in Figure 7.52. The temperature profile of the fabricated prototype absorber is measured at a non-absorbing frequency of 7.5

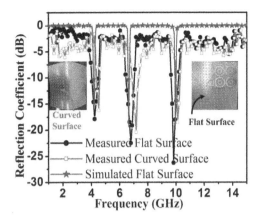

FIGURE 7.51
Measured reflection coefficients due to normal incidence of EM waves on flat and curved surfaces with a simulated flat surface.

Source: Copyright/used with permission of/courtesy of IEEE.

FIGURE 7.52
Measurement setup for temperature profile measurements.

Source: Copyright/used with permission of/courtesy of IEEE.

GHz first [Figure 7.53(a)]. A rise of about 1.5°C in the temperature of prototype absorber surface is recorded at the absorbing frequencies 4.19 GHz, 6.64 GHz and 9.95 GHz, as shown in Figure 7.53(b), (c) and (d), respectively [20]. It is also observed that the maximum EM energy is absorbed by the metallic portion of the prototype absorber.

A new approach for designing a multiband conformal ultrathin polarization-insensitive metamaterial absorber is successfully demonstrated. The proposed absorber is polarization-insensitive with wide angular stability up to 45° in azimuth and 30° in elevation plane. The absorption of EM waves by the structure is confirmed experimentally through lock-in IR thermography. A temperature increase of 1.5°C is observed when a microwave source is radiating at the absorbing frequencies as compared to the non-absorbing frequency [21, 22].

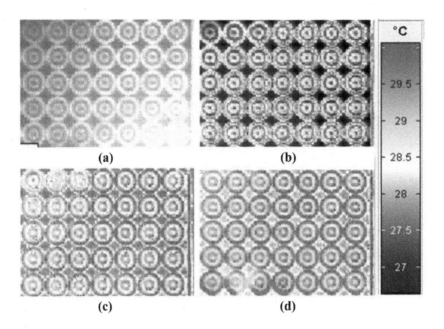

FIGURE 7.53
Measured temperature profile of proposed fabricated prototype triple-band absorber using lock-in IR thermography. Microwave source radiating at (a) 7.50 GHz; (b) 4.19 GHz; (c) 6.64 GHz and (d) 9.95 GHz.

Source: Copyright/used with permission of/courtesy of IEEE.

TABLE 7.1

Comparison Table of Proposed Absorbers with Similar Other Reported Absorbers

Ref.	Substrate Thickness (λ_0)	UC Size (λ_0)	Operating Bands (GHz)	Conformal
[22]	–	–	9.4, 11.7	No
[23]	$\lambda_0/138$	$0.216\lambda_0$	7.22, 14.55	Yes
[13]	$\lambda_0/126$	$0.253\lambda_0$	9.5, 15	Yes
Proposed Double-Band Absorber	$\lambda_0/142.65$	$0.167\lambda_0$	**4.19, 9.75**	**Yes**
[1]	$\lambda_0/92.24$	$0.12\lambda_0$	3.2, 9.45, 10.90	No
[11]	$\lambda_0/77$	$0.206\lambda_0$	4.11, 7.91, 10.13, 11.51	No
[15]	$\lambda_0/91$	$0.11\lambda_0$	4.19, 9.34, 11.48	No
Proposed Triple-Band Absorber	$\lambda_0/142.65$	$0.167\lambda_0$	**4.19, 6.64, 9.95**	**Yes**

7.8 Conclusion

In this chapter, the development of novel metamaterial absorbers is presented for S-, C-, X-, Ku- and Ka-band applications. The proposed absorbers are found to have thickness of $\lambda_0 / 34.50$, $\lambda_0 / 135.66$ and $\lambda_0 / 142.65$ [21], depending on the frequency of operation. The absorbers are experimentally characterized in an anechoic chamber. The absorber surface temperature is measured using lock-in IR thermography and temperature increase at absorbing frequencies is observed. Some of the absorbers are found to be conformal and polarization-insensitive for different angles of incidence of TE and TM waves with wide angular stability up to 45°. The proposed absorbers can be used for stealth technology in defence applications, radar applications and EMI/EMC reduction. The same prototype absorbers can be easily extended to multiband applications by introducing additional MCRRs.

References

1. Zhai, H., C. Zhan, Z. Li, and C. Liang, "A triple-band ultrathin metamaterial absorber with wide-angle and polarization stability," *IEEE Antennas and Wireless Propagation Letters*, vol. 14, pp. 241–244, 2015.
2. Ebrahimi, A. et al., "Second-order terahertz bandpass frequency selective surface with miniaturized elements," *IEEE Transactions on Terahertz Science and Technology*, vol. 5, no. 5, pp. 761–769, Sept. 2015.
3. Chen, J., Z. Hu, S. Wang, X. Huang, and M. Liu, "A triple-band, polarization- and incident angle-independent microwave metamaterial absorber with interference theory," *European Physical Journal B*, vol. 89, no. 1, pp. 143–152, 2016.
4. Yoo, M., H. K. Kim, and S. Lim, "Angular- and polarization-insensitive metamaterial absorber using subwavelength unit cell in multilayer technology," *IEEE Antennas and Wireless Propagation Letters*, vol. 15, pp. 414–417, 2016.
5. Fan, Y. et al., "An active wideband and wide-angle electromagnetic absorber at microwave frequencies," *IEEE Antennas and Wireless Propagation Letters*, vol. 15, pp. 1913–1916, 2016.
6. Liu, Xiaoming, Chuwen Lan, Bo Li, Qian Zhao, and Ji Zhou, "Dual band metamaterial perfect absorber based on artificial dielectric 'molecules'," *Scientific Reports*, vol. 6, 2016.
7. Hakla, N., S. Ghosh, K. V. Srivastava, and A. Shukla, "A dual-band conformal metamaterial absorber for curved surface," *2016 URSI International Symposium on Electromagnetic Theory (EMTS)*, Espoo, pp. 771–774, 2016.

8. Lin, X. Q., P. Mei, P. C. Zhang, Z. Z. D. Chen, and Y. Fan, "Development of a resistor-loaded ultrawideband absorber with antenna reciprocity," *IEEE Transactions on Antennas and Propagation*, vol. 64, no. 11, pp. 4910–4913, Nov. 2016.

9. Yoo, M., H. K. Kim, and S. Lim, "Angular- and polarization-insensitive metamaterial absorber using subwavelength unit cell in multilayer technology," *IEEE Antennas and Wireless Propagation Letters*, vol. 15, pp. 414–417, 2016.

10. Fan, Y. et al., "An active wideband and wide-angle electromagnetic absorber at microwave frequencies," *IEEE Antennas and Wireless Propagation Letters*, vol. 15, pp. 1913–1916, 2016.

11. Chaurasiya, D., S. Ghosh, S. Bhattacharyya, A. Bhattacharya, and K. V. Srivastava, "Compact multi-band polarisation-insensitive metamaterial absorber," *IET Microwaves, Antennas & Propagation*, vol. 10, no. 1, pp. 94–101, Jan. 9, 2016.

12. Sharma, Sameer Kumar et al., "Ultra-thin dual-band polarization-insensitive conformal metamaterial absorber," *Microwave and Optical Technology Letters*, vol. 59, no. 2, pp. 348–353, 2017.

13. Hasan, M. M., M. R. I. Faruque, and M. T. Islam, "A tri-band microwave perfect metamaterial absorber," *Microwave and Optical Technology Letters*, vol. 59, pp. 2302–2307, 2017.

14. Heydari, S., P. Jahangiri, A. Sharifi, F. B. Zarrabi, and A. Saee Arezomand, "Fractal broken cross with Jerusalem load absorber for multiband application with polarization independence," *Microwave and Optical Technology Letters*, vol. 59, pp. 1942–1947, 2017.

15. Mishra, N., D. K. Choudhary, R. Chowdhury, K. Kumari, and R. K. Chaudhary, "An investigation on compact ultra-thin triple band polarization independent metamaterial absorber for microwave frequency applications," *IEEE Access*, vol. 5, pp. 4370–4376, 2017.

16. Chaurasiya, D., S. Ghosh, S. Bhattacharyya, A. Bhattacharya, and K. V. Srivastava, "Compact multi-band polarisation-insensitive metamaterial absorber," *IET Microwaves, Antennas & Propagation*, vol. 10, no. 1, pp. 94–101, 2017.

17. Tak, J. and J. Choi, "A wearable metamaterial microwave absorber," *IEEE Antennas and Wireless Propagation Letters*, vol. 16, pp. 784–787, 2017.

18. Kurra, Lalithendra, Mahesh P. Abegaonkar, and Shiban K. Koul, "Equivalent circuit model of resonant EBG band stop filter," *IETE Journal of Research*, vol. 62, no. 1, pp 17–26, Jan. 2016.

19. Muzaffar, Khalid, Suneet Tuli, and Shiban Koul, "Determination of polarization of microwave signals by lock-in infrared thermography," *IETE Journal of Research*, pp. 123–130, Oct. 2015.

20. Singh, Amit K., Mahesh P. Abegaonkar, and Shiban K. Koul, "Penta band polarization insensitive metamaterial absorber for EMI/EMC reduction and defense applications," *2017 IEEE MTT-S International Microwave and RF Conference (IMaRC)*, Ahmedabad, 2017, pp. 1–5.

21. Singh, Amit K., Mahesh P. Abegaonkar, and Shiban K. Koul, "Dual- and triple-band polarization insensitive ultrathin conformal metamaterial absorber with wide angular stability," *IEEE Transactions on Electromagnetic Compatibility*, vol. 3, pp. 1–9, 2018.

22. Liu, Xiaoming, Chuwen Lan, Bo Li, Qian Zhao, and Ji Zhou, "Dual band metamaterial perfect absorber based on artificial dielectric 'molecules'," *Scientific Reports*, vol. 6, 2016.
23. Hakla, N., S. Ghosh, K. V. Srivastava, and A. Shukla, "A dual-band conformal metamaterial absorber for curved surface," *2016 URSI International Symposium on Electromagnetic Theory (EMTS)*, Espoo, pp. 771–774, 2016.

Index